Dictionary of New
INFORMATION TECHNOLOGY
ACRONYMS

Dictionary of New
INFORMATION TECHNOLOGY
ACRONYMS

M.Gordon A.Singleton C.Rickards

QA 76.15
G 67
1984

Kogan Page, London
Gale Research Company, Detroit

Copyright © Michael Gordon, Alan Singleton
and Clarence Rickards 1984
All rights reserved

First published in Great Britain in 1984 by
Kogan Page Limited, 120 Pentonville Road, London N1 9JN

British Library Cataloguing in Publication Data
Gordon, Michael
  Dictionary of new information technology acronyms
  1. Information science – Dictionaries
  I. Title   II. Singleton, Alan
  III. Rickards, Clarence
  020'3'21   Z1001

  ISBN 0-85038-572-5

First published in the United States of America,
possessions and Canada by Gale Research Company,
Book Tower, Detroit, Michigan 48226

ISBN 0-8103-4309-6

Printed in Great Britain by Nene Litho,
bound by Woolnough Bookbinding,
both of Wellingborough, Northamptonshire

Introduction

Information technology (IT) has been described as the merging of telecommunications, data processing and microelectronics into a force capable of revolutionizing our way of life. Indeed, the British Government's advisory body on the subject (ACARD) has stated unequivocally that IT will 'affect every household and occupation' and that the 'country's future trading performance will depend greatly on its ability to compete in world markets for products and services based on information technology and on the rapid effective application of such products and services by industry and commerce generally.'

While the importance of IT should not, therefore, be underestimated, public awareness of its capabilities is low. Indeed, a recent opinion poll showed that over 80 per cent of respondents had never heard of information technology. Clearly there is a lot of learning to be done. And if we are to survive in the new information society which appears to be upon us, then like it or not, we will have to cope with the technology – and that means understanding its jargon.

Some of this jargon is not new to IT; it has been used for some time in such fields as telecommunications, documentation, and computers. Other terms are, meanwhile, specific to IT. But all these types of jargon have one thing in common – the extensive use of acronyms and abbreviations. These terms aid communication between specialists in particular areas of IT but confuse people with only a general interest in IT, anyone coming across IT for the first time, or indeed a specialist from one field of IT looking into another such field. For it is impossible for any one person to maintain an awareness of all the acronyms and abbreviations used in every area of IT: the number of acronyms and abbreviations is so large and ever growing, and the same string of letters can have different expansions and meanings in different contexts.

Thus, when two of the present authors (Michael Gordon and Alan Singleton) worked on the compilation of a *Dictionary of New Information Technology*,* it became apparent that there was a

---

*Meadows, A.J., Gordon, M. and Singleton, A., Kogan Page, London, 1982.

need for a further dictionary dedicated to acronyms and abbreviations in the field of IT. And so they began work on this book.

The aim of the book is to give a single source which can be used to trace the meaning(s) of any given acronym or abbreviation relating to IT. This is done by presenting a list of such 'terms' in alphabetical order, with each term expanded and, where necessary, annotated. Some expansions are felt to make the meaning of a term self-evident, and in such cases no annotation has been included. Other terms have varying amounts of annotation, the detail depending on the specificity and clarity of the expansion, and the availability and accessibility of supplementary information elsewhere. Readers will vary in both their needs and expertise. Some readers will, therefore, find that particular annotations give all the information required to understand a term, while others will find that enough information is given to direct them to another source: for example, a directory of computer products or telecommunication services, a database guide or, of course, the *Dictionary of New Information Technology*.

So as to increase the economy of presentation, many entries are cross-referenced. This is done by italicizing a term in an expansion or annotation which is itself an acronym or abbreviation explained elsewhere in the book. A few other stylistic conventions are worth noting:
1. The terms (i.e. acronyms and abbreviations) appear in bold and are listed in alphabetical order, on a 'letter by letter' basis.
2. Terms with numbers are listed (i.e. alphabetized) as if spelled out; e.g. AD1 = ADONE. Roman numbers are treated in the same way.
3. Numbered, alternative expansions of a term which have an annotation applying to more than one such alternative are separated by a comma; (for example, 1. telephone, 2. telegram, telecommunications. In other cases, alternative expansions and annotations are separated by a full stop.
4. Databases, databanks and journals are all cited according to ISBN convention, i.e. with first letter only capitalized, unless proper names are included; e.g. *Journal of librarianship, Library Association record*. All are in italic.
5. In addition to such 'publications' only proper names, companies, persons, committees and countries, etc., are capitalized.

6. Where entries refer to the name of a commercial product, the company producing that product is given. However, many products are developed by a particular company and then the product name is adopted as a generic description of all products performing similar functions. In such cases, no company name is given. Some names are in translation between the specific and general application. The reader should therefore exercise caution in drawing inferences concerning ownership, trademarks and copyright of names.

As new acronyms appear continually, we plan further editions of this book in the future. We would, therefore, welcome any suggestions from readers for additions or alterations.

Abbreviations used for countries in the text:

FRG	Federal Republic of Germany
GDR	German Democratic Republic
UK	United Kingdom
US	United States (of America)
USSR	Union of Soviet Socialist Republics

A

A address.

A angström. Unit of length used for wavelengths.

AA 1. asset amount. 2. author affiliation. Searchable fields, Dialog *IRS*.

AAA *Astronomy and astrophysics abstracts.* Database (Astronomisches Rechen-Institut, GDR).

AAAB American Association of Architectural Bibliographers.

AAAC American Automatic Control Council.

AAB analytical abstract. Searchable field, *SDC*.

AACOBS Australian Advisory Council on Bibliographic Services.

AACR *Anglo-American cataloguing rules.*

AACR1 *Anglo-American cataloguing rules,* 1st edition.

AACR2 *Anglo-American cataloguing rules,* 2nd edition.

AACS asynchronous address communications systems.

AADE American Association of Dental Publishers.

AAEC Australian Atomic Energy Commission. Hosts *INIS* outside Europe.

AAED Academic American Encyclopedia. Database, *BRS*.

AAL absolute assembly language. Computer programming language.

AALL American Association of Law Librarians.

AAME American Association of Microprocessor Engineers.

AAP 1. Association of American Publishers. 2. associative array processor. 3. attached applications processor.

AAPL an array processing language. Computer programming language.

AARLMP Afro-American Resources and Library Manpower Project, Columbia University (US).

AARS automatic address recognition system. Computing.

AASL American Association of School Librarians of the *ALA*.

AAU address arithmetic unit. Computing.

AB 1. abstract. Used to define a searchable field in an *OLS*. 2. automated bibliography. 3. address bus. Computing.

ABA 1. American Bankers' Association. 2. American Booksellers' Association. 3. Antiquarian Booksellers' Association. 4. Australian Booksellers' Association.

ABACUS AB Atomenergi – computerized user-oriented services (Sweden).

ABANK *Annual state databank*, Kentucky. Databank *(KEIS).*

ABC 1. adaptable board computer, Signetics (US). 2. American Broadcasting Company. 3. Australian Broadcasting Commission. 4. automatic bandwidth control. 5. automatic block controller.

ABCA American Business Communication Association.

ABC-Clio American Bibliographical Center – Clio Inc. Database originator.

ABC POL SCI *A bibliography of contents – political science and government.* Database, *ABC-Clio.*

ABD 1. Association Belge de Documentation (also *BVD*). 2. Association des Bibliothécaires-Documentalistes de l'Institut d'Études Sociales de l'État = Association of librarians-documentalists of the state institute of social studies (Belgium).

ABE arithmetic building element. Computing.

ABEND Abnormal end. Computing.

9

ABES Association for Broadcast Engineering Standards (US).

ABF Association des Bibliothécaires Français = Association of French librarians.

ABI Association des Bibliothèques Internationales.

ABI/INFORM *Abstracted business information/information needs.* Database on business management and administration accessible via *BRS*, Lockheed and *SDC*. Also *INFORM*.

ABIISE Agrupacion de Bibliotecas para la Integracion de la Informacion Socio-Economica. Library Group for the Integration of Socio-Economic Information. Cooperative library group (Peru).

ABIPC *Abstract bulletin of the Institute of Paper Chemistry* (US).

ABIX *Australian business index.* Database on *AUSINET*.

ABL 1. architectural block diagram language. 2. Atlantic basic language. 3. *atlas* basic language. Computer programming languages. 4. automatic bootstrap loader. Computing.

ABLISS Association of British Library and Information Science Schools.

ABM 1. *Art bibliographies modern*. Database, *ABC-Clio*. 2. asynchronous balanced mode. Computing.

ABN Australian Bibliographic Network. Centralized shared cataloguing facility, National Library of Australia.

ABNO all but not only. Information retrieval system mode.

ABOA *Australian bibliography on agriculture.* Database on *AUSINET*.

ABOL adviser business oriented language. Computer programming language.

ABP 1. actual block processor, *IBM*. 2. advanced business processor, Datapoint Corporation (US).

ABPA Australian Book Publishers' Association.

ABS 1. abstract. 2. air baring surface. Part of *R/W* head in magnetic disc unit. 3. automatic bibliographic services.

ABSTI Advisory Board on Scientific and Technical Information (Canada).

ABSW Association of British Science Writers.

ABT Australian Broadcasting Tribunal. Carried out enquiry into cable *TV*.

ABU Asia-Pacific Broadcasting Union.

AC 1. accession number. Searchable field, Pergamon-Infoline and *SDC*. 2. accumulator. 3. acoustic coupler. 4. activity code. Searchable field, Dialog *IRS*. 5. address counter. Computing. 6. alternating current. 7. analog computer. 8. area code. Used in a variety of software contexts. 9. assignee code. 10. authority code. Searchable fields, Dialog *IRS*. 11. automatic checkout. 12. automatic computer.

ACA 1. American Communications Association. 2. asynchronous communications adaptor.

ACAC Associate Committee on Automatic Control, National Research Council (Canada).

ACAM augmented content addressed memory.

ACARD Advisory Council for Applied Research and Development. Advisory body to government (UK).

ACC 1. accumulator. 2. Amateur Computer Club (UK). 3. Association of Computer Clubs (UK). 4. asynchronous communications control(ler). 5. automatic colour control.

ACCAP autocoder-to-*COBOL* conversion aid program, IBM.

ACCC *Ad hoc* Committee for Competitive Communication (US).

ACCESS 1. aircraft communication electronic signalling system. 2. Algemene Classificatie Commisse voor de Overheids-Aadministratie. General classification committee for governmental administration (Netherlands). 3. architects' central constructional engineering surveying service, Greater London Council (UK). 4. architecture, construction and consulting engineers special services. Information service, Stanton Municipal Library (Australia). 5. Argonne code center exchange and storage system, US Atomic Energy Commission. 6. *Automated catalog of computer equipment and software* *systems*, US Army. 7. automatic computer controlled electronic scanning system.

ACCIS Advisory Committee for Coordination of Information Systems (UN).

ACCWP Acquisition, Cataloguing and Circulation Working Party of the Aslib Computer Applications Group (UK).

ACD 1. automatic call distributor, Datapoint (US). 2. automatic call distribution. Switching system.

ACE 1. animated computer education. 2. automatic checkout equipment. 3. automatic computer evaluation. 4. automatic computing engine. Early computer, *NPL*.

ACES automated code evaluation system.

ACF 1. advanced communications function, *IBM*. 2. *Les arrêts de la Cour Fédérale*. Legal database, Department of Justice (Canada).

ACG automatic code generator.

ACH automated clearing house. Part of an *EFTS*.

ACIA asynchronous communications interface adaptor. Provides data formatting and control for data communications.

ACIC Aeronautical Chart and Information Center, US Air Force.

ACID automated classification and interpretation of data. Computer programming language.

ACK affirmative acknowledge. Control character.

ACL 1. application control language. 2. Association for Computational Linguistics (US). 3. *Atlas* commercial language. 4. audit command language. Computer programming languages.

ACLS American Council of Learned Societies.

ACM 1. address calculation machine, *CIIHB*. 2. alterable control memory. 3. Association for Computing Machinery (US). 4. associative communication multiplexer.

ACMAC *ACM* Accreditation Committee.

ACMSC *ACM* Standards Committee.

ACN 1. accession number. 2. *Alternative catalog newsletter*, Johns Hopkins University (US).

ACO automatic call origination. Telecommunications.

ACOMPLIS a computerized London information service, Greater London Council (UK).

ACOPP abbreviated Cobol preprocessor.

ACORN 1. associative content retrieval network. Information system, A. D. Little Inc. (US). 2. Computer manufacturer (UK), not acronym.

ACOS a computer series, *NEC/NTIS*.

ACP 1. airline control program, *IBM*. 2. ancillary control processor, 3. arithmetic and control processor. Computing.

ACPA Association for Computer Programmers and Analysts (US).

ACP/TF airline control program/ transaction processing facility, *IBM*.

ACR 1. access control register. 2. alternate *CPU* recovery. 3. *American computer referal*. Databank (US).

ACRIT Advisory Committee for Research on Information Transfer. Committee reporting to Netherlands government.

ACRL Association of College and Research Libraries of the *ALA*.

ACRODABA acronym database. (US – cancelled 1973.)

ACROPOLI *Application de chronologie de politique internationale*. Database, Ministère des Affaires Étrangères (France).

ACS 1. Advanced Communications Service, *A T & T Corp*. 2. advanced computer series, Honeywell. 3. advanced computer system, *IBM*. 4. Altos Computer Systems (US). 5. American Chemical Society. 6. *Les arrêts de la Cour Suprême*. Legal database, Department of Justice (Canada). 7. Australian Computer Society. 8. automated communications set. 9. automatic checkout system. 10. auxiliary code storage. Computer memory.

ACSI Association Canadienne des Sciences de l'Information. Also called *CAIS*.

11

ACSYS accounting computer system, Burroughs (US).

ACT 1. advanced communications technology. 2. algebraic compiler and translator. 3. Alternative Community Telephone (Sydney, Australia). 4. Applied Computer Techniques. Computer manufacturer (UK). 5. automatic code translator.

ACTIS Auckland Commercial and Technical Information Service (New Zealand).

ACTRAN autocoder to *COBOL* translating.

ACTS automatic computer telex services.

ACTSU Association of Computer Time-Sharing Users (US).

ACU 1. arithmetic control unit. 2. Association of Computer Users (US). 3. automatic calling unit. Telecommunications device.

ACWS *All-Canada weekly summaries*. Legal database, Canadian Law Book Limited.

AD 1. analog to digital. 2. *ASTIA* document.

A/D analog/digital.

ADA 1. computer programming language. Not an acronym. 2. automatic data acquisition.

ADABAS adaptable database system. Database management system, used for cooperative cataloguing systems (US and FRG).

ADAC automated direct analog computer.

ADAM 1. a data management system. 2. advanced data management. Mitre Corporation (US). 3. automatic document abstracting method.

ADAPS automatic display and plotting system.

ADAPSO Association of Data Processing Organizations (US and Canada).

ADAS automatic data acquisition system.

ADAT automatic data accumulator and transfer.

ADB 1. book teleordering system (Denmark). 2. *Angendatenbank*. Databank on steel making (*VDEL*).

ADBS 1. advanced database system, *NEC/NTIS*. 2. Association des Documentalistes et des Bibliothécaires Spécialisé (France). Also *AFDBS*.

ADC 1. advise duration and charge. Telephone operator service (UK). 2. analog to digital conversion. Computing term.

ADCC asynchronous data communications channel.

ADCCP advanced data communications control procedure.

ADCIS Association for the Development of Computer-based Instructional Systems (US).

ADCU Association of Data Communications Users (US).

ADCVR analog-digital converter.

ADD address. Location in computer memory.

ADDA Australian Database Development Association.

ADDAR automatic digital data acquisition and recording.

ADDS Applied Digital Data Systems. US company, acquired by *NCR*.

ADE 1. address error. 2. advanced data entry. 3. automated debugging environment, *ADR* (US). 4. automated design engineering.

ADEPT automated direct entry packaging technique.

ADES 1. Association de la Documentation Économique et Sociale. Association for economic and social documentation (France). 2. automatic digital encoding system. Language used by US Naval Ordnance.

ADF automatic document feed.

ADI American Documentation Institute. Superseded by *ASIS*.

ADIC 1. Aktueel Dokumentie en Informatie-Centrum. Current information centre, Amsterdam Public Library (Netherlands). 2. analog to digital conversion.

ADIOS 1. analog digital input/output system. 2. automatic diagnosis input/output system. 3. automatic digital input/output system.

12

ADIS 1. Association for the Development of Instructional Systems (US). 2. automatic data interchange system. Teletype Corporation (US).

ADIT analog digital integration translator.

ADJ adjacent. Text search term.

ADL 1. Arthur D. Little. Major IT consultancy, also originates databases. US-based with European offices. 2. automatic data link.

ADLC advanced data link control. Link protocol.

ADLIB adaptive library management system. Library automation software.

ADM 1. adaptable data manager, Hitachi (Japan). 2. Advanced Microdevices Inc. (US).

ADMIRE 1. adaptive decision maker in an information retrieval environment. Stanford University (US). 2. automated diagnostic maintenance information retrieval system.

ADMIS automated data management information system.

ADO automatic dial-out.

ADONIS 1. automated document delivery over networked information service. Moribund scheme of large academic publishers to provide copies of journal articles from optical digital discs. 2. automatic digital online instrumentation system. 3. automatic document online information system. Document delivery system, Association of European Publishers.

ADOS advanced diskette operating system.

ADP 1. advanced data processing. 2. automatic data processing. 3. Association of Database Producers (UK).

ADP (Network Services) Advanced Data Processing Network Services. Host (UK).

ADPCM 1. adaptive differential pulse code modulation. 2. Association for Data Processing and Computer Management (US).

ADPE automatic data-processing equipment.

ADPP automatic data-processing program.

ADPS automatic data-processing system.

ADR Applied Data Research Inc. Software supplier – (US).

ADRAC automatic digital recording and control.

ADRES Army Data Retrieval System (US).

ADRS automatic document request service. Provided by *BLLD* via *BLAISE*.

ADRT analog data recorder transcriber.

ADS 1. advanced debugging system. 2. Advanced Digital Systems. Computer peripheral manufacturer (US). 3. Anker Data Systems. Library automation system producer. 4. automatic duplicating system, Itek Corporation (US).

ADSATIS Australian Defence Science and Technology Information System.

ADSTAR automatic document storage and retrieval.

ADSUP automatic data systems uniform practices. Programming language.

ADT 1. application(s) dedicated terminal. 2. asynchronous data transceiver. 3. automatic data translator. 4. autonomous data transfer.

ADTD Association of Data Terminal Distributors (US).

ADTS automated data and telecommunications service.

ADU 1. automatic data unit. 2. automatic dialling unit.

ADX 1. asymmetric data exchange. 2. automatic data exchange.

AE arithmetic element.

AEA 1. Aircraft Electronics Association (US). 2. American Electronics Association.

AEBIG *Aslib* Economics and Business Information Group.

AEC automatic exposure control, in reprographic systems.

AECT Association of Educational Communications and Technology (US).

AED 1. Advanced Electronics Design. Terminal manufacturer (UK). 2. automated engineering design. Form of *CAD*.

13

AEDP Association for Educational Data Processing (US).

AEDS Association for Educational Data Systems (US).

AEG 1. active element group. Storage call or logic gate in *IC*. 2. Allgemeine Elektrizitäts-Gesellschaft. AEG-Telefunken. *IT* manufacturer (FRG).

AEGIS agricultural ecological and geographical information system.

AEI 1. Associated Electrical Industries. UK company absorbed by *GEC*. 2. *Australian educational index*. Database (Australian Council for Educational Research).

AEL audit entry language, Burroughs (US).

AEM Association of Electronic Manufacturers (US).

AEON advanced electronics network. System for shared access to central resources (UK).

AEPS advanced electronic publishing system. *BPCC* system for integrating text capture, formatting, typesetting, electronic distribution and database preparation.

AEROS *Aerometric and emissions reporting system*. Databank, *EPA*.

AEROSAT aeronautical satellite, for air traffic control.

AES Audio Engineering Society (US).

AESC Aerospace and Electronic Systems Society (US).

AESI *Australian earth sciences information system*. Database on *AUSINET*.

AEWIS Army Electronic Warfare Information System (US).

AF 1. affiliation of first author. Used to define searchable field. 2. audio frequency.

AFC automatic frequency control.

AFCAL Association Française de Calcul = French computing association.

AFCC Association of Federal Communications Consulting Engineers (US).

AFCEA Armed Forces Communications and Electronics Association (US).

AFCET Association Française pour la Cybernétique Économique et Technique.

AFDAC Association Française pour la Documentation Automatique en Chimie. Information broker (France).

AFDBS Association Française des Documentalistes et des Bibliothécaires Spécialisés = French association of documentalists and special librarians, also known as ADBS.

AFEE *Association Française pour l'Étude des Eaux*. Database on the sea and the environment.

AFF automatic fast feed.

AFG analog function generator.

AFIP American Federation of Information Processing. Formerly *AFIPS*.

AFIPS American Federation of Information Processing Societies.

AFMDC Air Force Machinability Data Center, Metcut Research Association for the US Air Force.

AFNOR Association Française de Normalisation (French Society for Standards) Database/databank originator and operator (France).

AFO advanced file organisation.

AFP 1. Agence France Press. French news agency, supplies full text databases. 2. attached *FORTRAN* processor, Burroughs (US).

AFR 1. advanced fault recognition. 2. automatic field/format recognition. Facility within computer system.

AFRE *Australian financial review*. Database on *AUSINET*.

AFSARI automation for storage and retrieval of information. Information retrieval term.

AFSK audio frequency shift keying.

AFT automated funds transfer.

AFTN aeronautical fixed telecommunications network.

AG assets greater than. Searchable field (Dialog *IRS*).

AGC automatic gain control. Electronics.

AGDEX *Agricultural index*. Database and current awareness service, Edinburgh School of Agriculture (UK).

AGE 1. American Bibliography of Agricultural Economics. Database originator (US). 2. Asian Center for Geotechnical Engineering (Taiwan).

AGECON *American bibliography of agricultural economics*. Database (US).

AGE digest database, *AGE* (2).

AG-FIZ Arbeitsgemeinschaft der Fachinformationszentren. Working group of information centres (FRG).

AG-IDL Arbeitsgemeinschaft Information und Dokumentation Literaturversorgung. Working group on information and document supply (FRG).

AGLINE *Agriculture online*. Database, Doane Western Inc (US).

AGLINET Agricultural Libraries Information Network (UN).

AGNES algorithm for generating structural surrogates of English text. Computing procedure.

AGREP *Agricultural research projects*. Database of current agricultural research projects, *CEC*.

Agricola *Agricultural online access*. Database, US Department of Agriculture.

AGRINET inter-American system of agricultural information.

AGRIS *Agricultural information system*. Database on agriculture, Food and Agriculture Organisation (UN).

AGT 1. Alberta Government Telephones (Canada). 2. audiographic teleconference.

AH 1. acceptor handshake. Telecommunications. 2. analog hybrid. Computing.

AHAM Association of Home Appliance Manufacturers. Originator and databank on engineering and economics of manufacture (US).

AHB *Austrian historical bibliography*. Database (*IMD*).

AHCS advanced hybrid computer system.

AHL *American history and life*. Database, *ABC-Clio*.

AHONDA Ad Hoc Committee on New Directions of the Research and Technical Services Division of the *ALA*.

AI 1. *Alloys index*. File within *METADEX*. 2. artificial intelligence. Computing term. 3. *Artikkel indeks*. Database, *NSI*. 4. automatic input.

A and I abstracting and indexing.

AIAA American Institute of Aeronautics and Astronautics. Pioneered information systems in the 1960s and 1970s.

AIB analog input/output board.

AICA 1. Automatico Associazione Italiana per il Calculo. Computing association (Italy). 2. International Association for Analog Computing.

AICS Association of Independent Computer Specialists (UK).

AID 1. algebraic interpretive dialogue. Computing. 2. *Arbetslivets information och dokumentation*. Database on social sciences and public administration, Arbetslivscentrum (Sweden). 3. Association Internationale des Documentalistes et Techniciens de l'Information = International Association of Documentalists and Information Officers. 4. *Augmented index and digest*, Information Retrieval Ltd. (UK). 5. automatic information distribution. Information retrieval term.

AIDA analysis of interconnected decision areas. Management term.

AIDC automatic image density control. Toning technique used in some photocopiers.

AIDE automated integrative design engineering.

AIDI Associazione Italiana per la Documentazione e l'Informazione = Italian association for documentation and information.

AIDS 1. acoustic intelligence data system. 2. advanced interactive debugging system. 3. *Aerospace intelligence data system*, IBM (US). 4. Amdahl internally developed software, Amdahl (US). 5. automated information dissemination system.

6. automatic integrated debugging system.
7. Automation Instrument Data Service, Indata Ltd. (UK).

(Forest) AIDS *(Forest products) abstract information digest service.* Database, *FPRS.*

AIEE American Institute of Electrical Engineers. Merged with *IRE* to form *IEEE.*

AIET average instruction execution time. Computer parameter.

AIGA American Institute of Graphical Arts.

AIIM Association for Information and Image Management. Formerly National Micrographics Association (US).

AIKR artificial intelligence knowledge representation.

AIL array interconnection logic. Computing.

AIM 1. *Abridged index medicus.* Database on medical sciences, National Library of Medicine (US). 2. *Abstracts of instructional materials in vocational and technical education.* Database, now *AIM/ARM.*
3. access isolation mechanism. Computing.
4. advanced information manager, Fujitsu (Japan). 5. Association for Information Management. Secondary title of Aslib.
6. Association of Information Managers (US). 7. associative index method.

AIM/ARM *Abstracts of instructional materials and research materials in vocational and technical information.* Database, Center for Vocational Education, Ohio State University (US).

AIMC academic information management center (proposed) (US).

AIMLO auto-instructional media for library orientation. Reader education device, Colorado University Library (US).

AIMS automated information and management systems.

AIM-TWX *Abridged index medicus* – teletypewriter exchange network. Information network, *NLM* (US).

AIO analog input/output board.

AIOP automatic identification of outward dialling. Telecommunications.

AIOPI Association of Information Officers in the Pharmaceutical Industry (UK).

AIP 1. alphanumeric impact printer.
2. American Institute of Physics. Database originator.

AIRHPER Alberta information retrieval for health, physical education and recreation. Information retrieval system, University of Alberta (Canada).

AIRS 1. American Information Retrieval Service. Document delivery service for US government and other US organizations.
2. automatic image retrieval system.
3. automatic information retrieval system.

AIS 1. accounting information system.
2. Aeronautical Information Service, Civil Aviation Authority (UK). 3. analog input system. 4. automatic intercept system. Telecommunications.

AISC Association of Independent Software Companies.

AIT advanced information technology.

AJ Anderson Jacobson. Terminal manufacturer.

AKR address key register.

AKWIC author and keyword in context. Indexing system, see *KWIC.*

AL 1. assembly language. 2. assets less than. Searchable field, Dialog *IRS.*

ALA American Library Association.

ALADIN Automatisering landbouwkundige dokumentatie-en informatievespreiding in Nederland = automation of agricultural documentation and information in the Netherlands, Centre for Agricultural Publishing and Documentation (Netherlands).

ALA/ISAD American Library Association Information Science and Automation Division.

ALAP associative linear array processor. Computing.

ALARM anaesthesia literature abstracting retrieval method. American Society of Anesthesiologists.

ALAS automated literature alerting system. Current awareness system.

ALBIS Australian library-based information system, National Library of Australia.

ALBO *Albo degli avvocatie procuratori*. Legal databank, *CSC* (Italy).

ALC 1. adaptive logic circuit. 2. assembly language coding. Computing.

ALCAPP automatic list classification and profile production.

ALCC Association of London Computer Clubs.

ALCOM algebraic compiler.

ALCS Authors' Lending and Copyright Society (UK).

ALCU 1. arithmetic logic and control unit. Computing. 2. asynchronous line control unit. Telecommunications.

ALD 1. analog line driver. 2. asynchronous line driver.

ALEPH automatic library expendable program Hebrew University. Online library system, Hebrew University (Israel).

ALERT automatic linguistic extractor and retrieval technique.

ALF 1. application library file. Computing. 2. automatic line feed. Telecommunications.

ALFTRAN *ALGOL* to *FORTRAN* translator.

ALGO algebraic compiler, based on *ALGOL*.

ALGOL algorithmic language. Computer programming language.

ALI 1. asynchronous line interface. Telecommunications. 2. automated logic implementation. Computing.

ALIS automated library information system of the Technological Library of Denmark, also a database on Datacentralen, *DTB*.

ALIT *Australian literature*. Database, Murdoch University and *WAIT*.

ALLC Association for Literary and Linguistic Computing (US).

ALM 1. assembler language for *MULTICS*. 2. asynchronous line module. 3. asynchronous line multiplexor. Telecommunications.

ALMA alphanumeric code for music analysis. Input code for music notation.

ALMS analytic language manipulation system.

ALOFT Airborn Light Optical Fiber Network (US).

ALP 1. arithmetic logic processor. 2. assembly language processing. 3. automated language processing. Computing. 4. automated learning process.

ALPAC Automated Language Processing Advisory Committee of the National Academy of Sciences (US).

ALPHA automated literature processing handling and analysis, US Army.

ALPS 1. advanced linear programming system. Operational research technique. 2. associative logic parallel system. Computing. 3. automated library processing services, *SDC*.

ALPSP Association of Learned and Professional Society Publishers (UK).

ALS 1. arithmetic logic section. Computing. 2. Automated Library Systems Ltd. Manufacturer of automated circulation control systems (UK).

ALSA(s) area library service authority/authorities (US).

ALTRAN 1. algebraic translator. *FORTRAN* extension. 2. assembly language translator, Xerox.

ALTS automated library technical services, Los Angeles Public Library (US).

ALU 1. arithmetic and logic unit. Computing. 2. asynchronous line unit. Telecommunications.

ALWIN algorithmic Wiswesser notation. Chemical information retrieval technique.

AM 1. address modifier. 2. address(ing) mode. Computing. 3. amount of grant or contract. Searchable field, Dialog *IRS*. 4. amplitude modulation. Telecommunications. 5. associative memory. Computing. 6. asynchronous modem. Telecommunications. 7. auxiliary memory. Computing.

AMA 1. associative memory address. 2. associative memory array. Computing. 3. automatic message accounting. Telecommunications.

AMAIS *Agricultural materials analysis information service*, Laboratory of the Government Chemist (UK).

AMA/NET American Medical Association Network.

AMC 1. automatic message counting. 2. autonomous multiplexer channel. Telecommunications.

AMD Advanced Micro Devices. Computer manufacturer.

AMDAC Amdahl Diagnostic Resistance Center (US).

AMDS Australian *MARC* Distribution Service. Centralized cataloguing service (Australia).

AME automatic microfiche editor.

AMFIS automatic microfilm information system.

AMI 1. *Advertising and market intelligence*. Databank, *NYTIS*. 2. alternative mark inversion signal. Telecommunications. 3. American Microsystems Inc.

AMIS Agricultural Management Information System, Centre for Information and Documentation, EEC.

AML array machine language. Computing.

AMLC asynchronous multiline controller. Telecommunications.

AMLCC asynchronous multiline communications coupler. Telecommunications.

AMM 1. additional memory module. 2. analog monitor module. Computing.

AMNIPS adaptive man/machine non-numeric information processing system, *IBM*.

AMOS 1. Alpha Microsystems operating system, Alpha Microsystems (US). 2. associative memory organization systems. Compiler.

AMP associative memory processor. Computing.

AMPA American Medical Publishers Association.

AMPL 1. advanced microprocessor programming language. 2. Advanced Microprocessor Prototyping Laboratory, Texas Instruments (US).

AMPS automatic message processing system. Telecommunications.

AMR 1. automatic message registering. 2. automatic message routing. Telecommunications.

AMRA American Medical Record Association.

AMRS *Australian machine readable cataloguing record service*. Database, *NLA*.

AMS 1. Advanced Memory Systems (US). 2. American Mathematical Society. 3. asymmetric multiprocessing system, *IBM*.

AMTC Association for Mechanical Translation and Computation Linguistics (US).

AMTD automatic magnetic tape dissemination. Dissemination service, Defense Documentation Center (US).

AMU Association of Minicomputer Users (US).

AMVER Atlantic merchant vessel alert. Computer-based disaster alert system.

AMWA American Medical Writers' Association.

AN 1. abstract number. Searchable field, Dialog *IRS*. 2. accession number. Searchable field, Dialog and *SDC*. 3. agency name. Searchable field, Dialog *IRS*. 4. alphanumeric. 5. assignee name. Searchable field, Dialog *IRS*.

ANA 1. Article Numbering Association. Concerned with machine-readable labelling of retail merchandise (UK). 2. automatic number analysis.

ANACOM analog computer.

ANACONDA *Ad hoc* Committee to Work with Activities Committee On New Directions for *ALA* Activities (US).

ANATRAN analog translator.

ANB *Australian national bibliography*. Database, *NLA*.

ANCE Automated News Clipping Indexing and Retrieval System, Image Systems (UK).

ANEDA Association Nationale d'Études pour la Documentation Automatique = National association for studies in automatic documentation (France).

ANI automatic number identification. Telecommunications.

ANLES analog switch. Microprocessor.

ANPA American Newspaper Publishers' Association.

ANRIC Annual National Information Retrieval Colloquium (US).

ANS American National Standards.

ANSA 1. advanced network system architecture, *NEC/NTIS*. 2. *Automatic new structure alert*. Database on chemistry.

ANSCR alpha-numeric system for classification of recordings.

ANSI American National Standards Institute.

ANSLICS Aberdeen and North Scotland Library and Information Cooperative Services.

ANSTI African Network for Scientific and Technological Institutions, set up by *UNESCO*.

Antiope l'aquisition numérique et télévisualisation d'images organisées en pages d'écriture. Videotex system (France).

Antiope-Didon teletext system, *Antiope*.

Antiope-Titan viewdata system, *Antiope*.

ANTS *ARPA* network terminal system.

AOC automatic output control.

AOCU 1. arithmetic output control unit. 2. associative output control unit. Computing.

AOD arithmetic output data. Computing.

AOF advanced operating facility, Computer Technology Inc. (US).

AOI *Accent on information*. Medical databank, Cheever Publishing Inc.

AOIPS atmospheric and oceanic information processing system. Satellite image enhancing system.

AOL application oriented language. Computing.

AOS 1. advanced operating system, Data General (US). 2. algebraic operating system, Texas Instruments (US). 3. author organization source. Searchable field, *SDC*.

AOSI *Alberta oil sands index*. Database, Alberta Oil Sands Information Center (Canada).

AOS/VS advanced operating system/virtual storage, Data General (US).

AOU 1. arithmetic output unit. 2. associative output unit. Computing.

AP 1. application program. Computing. 2. approach. Searchable field, Dialog *IRS*. 3. arithmetic processor. 4. array processor. Computing. 5. Associated Press. News agency and wire service. 6. associative processor. 7. attached processor. Computing.

APA American Psychological Association. Database originator.

APACE Aldermaston project for the application of computers to engineering. *UKAEA*.

APAIS *Australian public affairs information service*. Databank, *NLA*.

APAL array processor assembly language. Computing.

APAM array processor access method. Computing.

APAR automatic processing and recording. Compiler.

APC 1. advanced personal computer. 2. associative processor control. Computing. 3. automatic phase control.

APCS associative processor computer systems.

APE(x)C all purpose electronic x computer. Early computer, Birkbeck College (UK).

APG automatic priority group, Fujitsu (Japan).

API 1. American Petroleum Institute. Database originator. 2. Auerbach power index. Benchmark used in *Auerbach computer technology reports*.

APILIT *American Petroleum Institute Literature*. Database on petroleum and petrochemicals, *API*.

APIPAT *American Petroleum Institute patents*. Database on petroleum and petrochemical patents, *API*.

APK amplitude phase keyed. Telecommunications.

APL 1. a programming language. High level language. 2. algorithmic programming language. 3. associative programming language. 4. average picture level.

APM automatic predictive maintenance.

APOLLO article procurement with online local ordering. Document delivery system.

APP associative parallel processor. Computing.

APPECS adaptive pattern perceiving electronic computer system.

APPLE associative processor programming language evaluation.

APR alternate path reentry, Fujitsu (Japan).

APRIS Alcoa picturephone remote information system. Telecommunications device, *A T & T* (US).

APS 1. array processor software, Data General (US). 2. attached processor for speech, *IBM*. 3. Augustan prose sample. Machine readable selection of English prose. 4. automatic page search. Document retrieval aid, Imtec Company. 5. auxiliary program storage.

APSE *ADA* programming support environment. Software engineering project (UK).

APSK amplitude and phase shift keying.

APSP array processor subroutine package.

APSS automated program support system.

APT 1. automatic picture transmission. 2. automatic programming tool. 3. automatically programmed tools. *CNC* software.

APTIC Air Pollution Technical Information Center. Database coordinated by *EPA* and the Franklin Institute (US).

APU 1. analog processing unit. 2. arithmetic processing unit. 3. asynchronous processing unit. 4. auxiliary processing unit.

AQL acceptance quality level. Engineering term.

AR 1. accumulator register. 2. address register. 3. arithmetic register. 4. associative register. Computing. 5. authority record. Searchable field, Dialog *IRS*.

ARC 1. Atlantic Research Corporation (US). 2. attended resource computer, Datapoint (US). 3. automatic relay calculator. Early computer, Birkbeck College (UK).

ARCS 1. advanced reconfigurable computer system. 2. automated ring code search. Information retrieval method, chemical structures.

ARD answering recording and dialling.

ARDIC Association pour la Recherche et le Développement en Informatique Chimique. Host and database originator (France).

ARDIS Army Research and Development Information System (US).

ARF automatic report feature.

ARGOS *Accumulation régionale de données géographiques organisées pour la statistique*. Regional collection of geographical statistical data. Databank covering census data, *INSEE*.

ARGUS automatic routine generating and updating system. Compiler.

ARI *Australian road index*. Database originated and operated by the Australian Road Research Board.

ARIST 1. Les Agences Régionales de l'Information Scientifique et Technique (France). 2. *Annual review of information science and technology*. Publication, Encyclopedia Britannica for *ASIS*.

ARISTOTLE Annual Review of Information and Symposium on the Technology of Training and Learning and Education. Annual symposium (US).

ARL Association of Research Libraries (US).

ARM 1. asynchronous response mode. Computing. 2. automated route management. Business management system. 3. availability, reliability and maintainability. Computer performance.

ARMA American Records Management Association.

ARMS automated record management system, Tyne & Wear County Council (UK).

ARO 1. after receipt of order. 2. automatic recovery option, *NCR*.

AROM alterable read-only memory.

AROS alterable read-only operating system.

ARPA Advanced Research Projects Agency, US Department of Defense. Also sometimes refers to *ARPANET*.

ARPANET Advanced Research Projects Agency Network. Computer network (US). See also *ARPA*.

ARQ automatic request for correction.

ARR address record register.

ARS audio response system.

ARSC automatic resolution selection control.

ART 1. automated request transmission. Ordering service offered by *BLLD* (UK). 2. automatic reporting telephone.

ARTEMIS automatic retrieval of text through European multipurpose information services. Document delivery system.

ARU 1. address recognition unit. 2. arithmetical unit. Computing. 3. audio response unit. Telecommunications.

AS 1. advanced systems, *NAS*. 2. agency state. Searchable field, Dialog *IRS*. 3. auxiliary storage. Computing.

ASA American Standards Association. Now *ANSI*.

ASBU Arab States Broadcasting Union.

ASC 1. advanced scientific computer, *TI*. 2. American Satellite Corporation. 3. American Society for Cybernetics.

ASCA automatic subject citation alert. Current awareness service.

ASCC Aeronautical Satellite Communications Center (US).

ASCE American Society of Civil Engineers. Originator and database.

ASCENT assembly system for central processor.

ASCII American standard code for information interchange. Code for data and text transmission.

ASCP automatic system checkout program.

ASCT address space control task, Fujitsu (Japan).

ASCU Association of Small Computer Users (US).

ASCUE Association of Small Computer Users in Education (US).

ASD Association Suisse de Documentation. Also *SVD*.

ASDI automated selective dissemination of information.

ASES automated software evaluation system.

ASF automatic sheet feeder.

ASFA *Aquatic sciences and fisheries abstracts*. Database, *FAO*, *IRL* and *IOC*.

ASFIS aquatic sciences and fisheries information system.

ASI 1. Addressing Systems International. Manufacturer of mailing list systems (UK). 2. *American statistics index*. Databank, Congressional Information Service (US). 3. articulated subject index. 4. asynchronous serial interface.

ASID address space identifier. Computing.

ASIDIC Association of Scientific Information Dissemination Centers (US).

ASII *Australia science index*. Database on *AUSINET*.

ASIN *Agricultural service information network*. National Agricultural Library (US).

ASIS American Society for Information Science.

ASIST advanced scientific instruments symbolic translator. Assembly program.

ASK 1. access to sources of knowledge. Referral system, now terminated (Canada). 2. amplitude shift keying. Telecommunications. 3. anomalous state of knowledge. Term used in artificial intelligence and concept experimentally used in information systems. 4. Applied Systems Knowledge Ltd. Software publisher (UK).

ASL Association for Symbolic Logic (US).

ASLT advanced solid logic technology.

ASM 1. American Society for Metals. Database originator. 2. assembler, Sperry Univac (US). 3. Association for Systems Management (US).

ASMI Agudat Ha-Sifriyot Ha-Meyuhadot Imerkeze Ha-Meda Beyisrael. Israel society of special libraries and information centres. Also *ISLIC*.

ASP 1. associative storing processor. 2. associative structures package. 3. attached support processor. 4. automatic schedule procedures. Computing.

ASPER assembly program for peripheral processes. Assembly program.

ASPI asynchronous synchronous programmable interface. Computing.

ASPIC author's standard pre-press interface code. Markup code for word processing/typesetter interface.

ASR 1. active status register. 2. address shift register. Computing. 3. answer, send and receive. 4. automatic send receive. Telecommunications. 5. automatic speech recognition.

ASS 1. analog switching subsystem. Telecommunications. 2. assembler, Siemens (FRG).

ASSASSIN agricultural system for storage and subsequent selection of information. Information retrieval system (ICI, UK).

ASSIRIS assembler pour *IRIS*. Computer series, *CIIHB*.

ASSISTENT automated information system for science and technology (USSR).

AST anti-sidetone. Telecommunications.

ASTED Inc. (Association) pour l'Avancement des Sciences et des Techniques de la Documentation (Canada).

ASTIA Armed Services Technical Information Agency, now Defense Documentation Center (US).

ASTINFO Scientific and Technical Information in Asia. Regional network for information exchange, *UNESCO*.

ASTM infrared data bases *American Society for Testing and Materials infrared data bases*. Databank of condensed spectra with chemical descriptors, Sadtler Research Labs (US).

ASU 1. apparatus slide-in unit. 2. automatic switching unit. Telecommunications.

ASV automatic self verification. Computing.

ASVIP American standard vocabulary for information processing.

ASVS automatic signature verification system.

AT 1. address translator. Computing. 2. article type. 3. asset type. Searchable fields, Dialog *IRS*. 4. automatic translation.

ATA 1. *Abstracts on tropical agriculture*. Database, Royal Tropical Institute (Netherlands). 2. American Translators Association. 3. asynchronous terminal adaptor. Telecommunications.

AT&T American Telephone and Telegraph Company.

ATB 1. access type bits. Computing. 2. all trunks busy. Telecommunications.

ATDM asynchronous time division multiplexer. Telecommunications.

ATE 1. artificial traffic equipment. Telecommunications. 2. automatic telephone exchange. 3. automatic test equipment.

ATF automatic test formatter.

AT/L *Advanced technology/libraries*. Newsletter, Knowledge Industry Publications.

ATLA American Theological Library Association. Database originator (US).

ATLAS 1. abbreviated test language for avionics systems. 2. automatic tabulating, listing and sorting system. Word processing software package.

ATLS *Australian transport literature information system*. Database on *AUSINET*.

ATM automated teller machine. Electronic banking.

ATOM automatic transmission of mail. Early electronic mail system (US).

ATR automatic traffic recorder. Telecommunications.

ATS 1. administrative terminal system. Early text handling system. 2. applications technology satellite. Communications satellite. 3. audio test set.

ATSP *Air transport statistical programme*. Databank, *ICAA*.

ATSS automatic telephone switching system.

ATSU Association of Time Sharing Users (US).

ATU 1. address translation unit. 2. autonomous transfer unit. Computer hardware.

AU 1. arithmetic unit. Computing. 2. author. Searchable field, *BLAISE, Dialog* and *SDC*.

AUCBE Advisory Unit for Computer Based Education (UK).

AUDACIOUS automatic direct access to information with online *UDC* system. Information retrieval system, American Institute of Physics.

AUDREY audio reply.

AUI attachment unit interface. Computing.

AUNT automatic universal translator.

AUSINET Australian information network.

AUSMARC Australian *MARC*.

AUSTPAC Australian packet switching service. Telecommunications.

AUT author. Field, Pergamon-Infoline.

AUTOCITE *Automated citation testing service*. Legal database, Lawyers Cooperative Publishing Co. (US).

AUTODIN automatic digital network, Department of Defense (US).

AUTOMAP automatic machining program.

AUTOMEX automatic message exchange service.

AUTOPIC automatic personal identification code, *IBM*.

AUTOPROMPT automatic programming of machine tools, *IBM*.

AUTOPSY automatic operating system.

AUTOSDI automatic *SDI*. Service from *BLAISE*.

AUTOSEVOCOM automatic secure voice communications.

AUTOVON automatic voice network, Department of Defense (US).

AV 1. audiovisual. 2. availability. Searchable field, *SDC*.

AVA absolute virtual address. Computing.

AVC automatic volume control.

AVD alternate voice and data. Telecommunications.

AVG average.

AVHRR advanced very high resolution radiometer. Used to 'photograph' earth from satellites.

AVIP Association of Viewdata Information Providers (UK). Also *BVIP*.

AVLINE *Audiovisual online*. Database on audiovisual aids in health science teaching, National Library of Medicine (US).

AVMARC *Audiovisual machine readable catalogue*. Database on audiovisual materials, *BLAISE*.

AVP attached virtual processor.

AVR automatic voice recognition.

AW Awardee. Searchable field, *SDC*.

AWC Association for Women in Computing.

AWGN additive voice gaussian noise. Telecommunications.

AWRE Atomic Weapons Research Establishment. Host (UK).

AY accession number year. Searchable field, Pergamon-Infoline and *SDC*.

AZ azimuth.

B

B 1. bit. 2. buffer. 3. byte.

BA 1. binary add. Computing. 2. *Biological abstracts*. Database (US).

BABT British Approval Board for Telecommunications.

BACE basic automatic checkout equipment.

BACS 1. Banks Automated Clearing System (UK). 2. bibliographic access and control system, Washington University School of Medicine, St Louis (US).

BADADUQ *Banque de Données à accès direct de l'Université du Québec*. Database of the holdings of the University of Quebec (Canada).

BAG bibliographic and grouping system. Program package for iconography (US).

BAL 1. basic assembler language, Sperry Univac (US). 2. business application language.

BALGOL Burroughs algebraic compiler.

BALIS *Bayerisches landwirtschaftliches Informationssystem*. Agricultural databank, Bayerisches Staatministerium für Ehrnährung, Landwirtschaft und Forsten (FRG).

BALLOTS bibliographic automation of large library operations using a time-sharing system. Cataloguing network, Stanford University (US). Now *RLIN*.

BAM 1. basic access method. 2. block access method. Computing. 3. Bundesanstalt für Materialprüfung. Database originator and operator (FRG).

BAMS *Bulletin of the American Mathematical Society*.

BAP basic assembler program.

BA Previews *Biological abstracts previews*. Database, *BIOSIS*.

BAR 1. bar address register. 2. buffer address register. Computing.

BARD *British database on research for aids for the disabled*, *HPRU*.

BAR Network/Spot TV *Broadcast advertisers' reports network/spot television*. Databank on TV advertising networks, Broadcast Advertisers Report (US).

BAS Bibliotheks automatisierung-system. Online cataloguing system (FRG).

BASH Booksellers' Association Service House. UK promotional organization.

BASIC 1. Basel Information Centre for Chemistry. Database operator serving a confederation of drug companies. 2. basic algebraic symbolic interpretive compiler. 3. basic automatic stored instruction computer. 4. battle area surveillance and integrated communications. 5. beginners algebraic symbolic interpretive compiler. 6. beginners all-purpose symbolic interaction code. Computer programming languages. 7. biological abstracts subjects in context, *BIOSIS*.

BASIS 1. bank automated service information system. 2. Battelle automatic search information system, Battelle Memorial Institute (US). 3. *Bulletin of the American Society for Information Science*.

BASIS-E Bibliothekarisch-analytisches system zur informations speicherung-erschleissung. Library analytical system for information storage/retrieval (FRG).

BASt *Bundesanstalt für Strassenwesen*. Transport database originator (FRG).

BASYS basic system.

BAT *Biological abstracts on tape*. *BIOSIS* service.

BATAB Baker and Taylor automated buying. Teleordering system (US).

BATS basic additional teleprocessing support. Computing.

BAUFO *Bauforschungsprojekte und berichte*. Databank of civil engineering research, *IRB*.

BAVIP British Association of Viewdata Information Providers.

BB base band.

BBC British Broadcasting Corporation.

BBK Bibliotechno-bibliograficheskaya klassifikatsiya. Library bibliographical classification (USSR).

BBL basic business language.

BBS bulletin board system. Personal computer message network system.

BC 1. Bibliographic classification. Library classification system, also called Bliss classification. 2. binary code. 3. binary counter. 4. biosystematic code. Searchable field in *OLS*. 5. branch city. Searchable field, Dialog *IRS*. 6. *Broadcasting stations*. Databank, *ITU*.

BCAVM *British catalogue of audiovisual materials*, British Library.

BCC 1. block check character. Telecommunications. 2. blocked calls cleared. Telecommunications, also *LCC*.

BCCL Birkbeck College Computation Laboratory (UK).

BCD 1. binary coded decimal. 2. *Business conditions digest*. Databank, *SDC*.

BCDC binary coded decimal counters.

BCDIC binary coded decimal interchange code.

BCI binary coded information.

BCIP Belgian Centre for Information Processing.

BCH 1. bids per circuit per hour. 2. blocked calls held. Telecommunications.

BCL 1. bar coded label. 2. Burroughs common language, Burroughs (US). 3. Business Computers Ltd. (UK).

BCLN British Columbia Library Network.

BCM *British catalogue of music*, *BL*.

BCML Burroughs current mode logic. Burroughs (US).

BCN Biomedical Communication Network (US).

BCO binary coded octal.

BCP 1. bisynchronous communications processor, *HP*. 2. Burroughs control program, Burroughs (US).

BCPA British Copyright Protection Association.

BCPL basic combined programming language.

BCPS basic call processing system. Telecommunications.

BCR 1. Bibliographic Center for Research, Denver, Colorado (US). 2. buffer control register.

BCS 1. binary communications: synchronous. Also *BYSINC*. 2. Biomedical Computing Society (US). 3. bridge control system. Telecommunications. 4. British Computer Society.

BCSI Biometric Computer Service Inc.

BCT between commands testing. Computing.

BCU binary counting unit.

BCW buffer control word.

BD 1. baud. Unit of transmission speed. 2. binary decoder. 3. binary divide. 4. binary to decimal.

BDAM basic direct access method, *IBM*.

BDC binary decimal counter.

BDCB buffered data/control bus.

BDD binary to decimal decoder.

BDDT *Banque de données en toxicologie*. Databank covering animal tests with toxic substances, *INSERM*.

BDIAC Battelle Defender Information Analysis Center, Battelle Memorial Institute (US). Now called *STOIAC*.

BDIC binary coded decimal interchange code.

BDLC Burroughs data link control, Burroughs (US).

BDN Bell Data Network.

BDOS batch disc operating system.

BDR 1. bi-duplexed redundancy. Telecommunications. 2. binary dump routine.

BDS bulk data switching.

BDST *Banque de données synthétisées en toxicologie.* Databank covering properties of toxic substances, *INSERM* (France).

BDT *Banque de données toxicologiques.* Databank on toxicology, Equipe de Recherche sur les Banques de Données (France).

BDTS buffered data transmission simulator.

BDU 1. *Banque de données urbaines (de Paris et de la Région Parisienne).* Databank on the people and buildings of Paris. Atélier Parisien d'Urbanisme (France). 2. basic display unit.

BEAB British Electrotechnical Approvals Board.

BEAMA British Electrical and Allied Manufacturers Association.

BEAMOS beam accessed metal oxide semiconductor. Memory technology.

BEAR Berkeley elites automated retrieval. Information retrieval system, University of California at Berkeley (US).

BEAST basic experimental automatic syntactic translator, Bunker Ramo Corporation (US).

BEBA *Bilingual education bibliographic abstracts.* Database, *NCBE*.

BECS basic error control system.

BECTIS Bell College technical information service, Hamilton (UK).

BED *Business equipment digest* (UK).

BEEF business and engineering enriched *FORTRAN*. Computer programming language.

BEFAP Bell Laboratories *FORTRAN* assembly program.

BEI *British education index.* Database, *BL*.

BEL bell. Control character, *ASCII*.

BELINDIS Belgian Information Dissemination Service. Host.

BELLREL Bell Laboratories library real time loan. Library loan system, Bell Telephone Laboratories (US).

BELLTIP library online acquisitions and cataloguing system, Bell Telephone Laboratories (US).

BEMA Business Equipment Manufacturers Association (US).

BER bit error rate. Data communications.

BERM bit error rate monitor.

BERNET Berliner Rechner Netz. Computer network (FRG).

BERT bit error rate test.

BES basic executive system, Honeywell (US).

BESSY Bestell system. Teleordering system developed by book wholesalers *KNO* (FRG).

BEST business *EDP* systems technique, *NCR*.

BETA Business Equipment Trade Association (UK).

BEU basic encoding unit.

BEX broadband exchange. Telecommunications system, Western Union (US).

BF base file.

B factor penalty factor, *VNL*.

BFAS basic file access system.

BFD basic floppy disc.

BFL busy flash. Telecommunications.

BFO beat frequency oscillator.

BGI *Bureau Gravimétrique International.* Databank on earth sciences, *BRGM*.

BGRAF basic graphics software, Magnavox (US).

BH binary to hexadecimal.

BI 1. basic index. Searchable field, *SDC*. 2. batch input. 3. *Brannvern indeks.* Databank, *NBF*.

BIAM PA *Banque d'information automatisée sur les médicaments principes actifs.* Databank on active ingredients of drugs, Association BIAM (France).

BIAM S *Banque d'information automatisée sur les médicaments spécialités.* Databank on pharmaceutical products, Association BIAM (France).

BIAS 1. Bibliotheks Ausleihverwaltungsystem. Library circulation system (FRG). 2. biomedical instrumentation advisory service, Clinical Research Centre (UK).

BIBDES bibliographic data entry system. *MARC* format cataloguing software, *BLAISE* (UK).

BIBLIO *Bibliografia (nazionale Italiano)*. Database of Italian books, Biblioteca Nazionale Centrale (Italy).

BIBLIODATA *Bibliographische Datenbank (der Deutschen Bibliothek)*. Database of books and periodicals published in the FRG, *GID*.

BIBLIOFILE small library automation system, Information Planning Associates Inc. (US).

BIBLIOS book inventory building library information oriented system. Orange County Public Library, California (US).

BIC 1. Biodeterioration Information Centre. Database originator and operator, University of Aston in Birmingham (UK). 2. bus interface controller.

BICEPT book indexing with context and entry point from text. Indexing method.

BID 1. *Bibliografia di informatica e diritto*. Bibliography of legal/rights information, *CSC*. 2. Bibliothekswesen, Informations- und Dokumentationswesen als zusammengehörender Bereich. Library and information science (FRG).

BIDAP bibliographic data processing program. Information retrieval software package.

BIDS Burroughs information display systems, Burroughs (US).

BIF *IF* bandwidth.

BIIT Bureau International d'Information sur les Télécommunications. Also the name of its database (Switzerland).

BIKAS das Bibliotheks-katalog System. Automated cataloguing system (FRG).

BIL block input length.

BIM 1. beginning of information marker. 2. branch if multiplexer. 3. bus interface module.

BIO buffered input/output.

BIONIC biological and electronic.

BIOP buffer input/output processor.

BIOR Business input/output return. Univac system.

BIOS basic input/output system.

BIOSIS *Biosciences information service*. Database and originator (US).

BIP 1. basic interpreter package, for *BASIC* language. 2. binary image processor. Computing. 3. *Books in print*. Database, R. R. Bowker (US).

BIPA *Banque d'information politique et d'actualité*. Database on political speeches, press cuttings etc, Télésystèmes-Questel (France).

BIPAC Bibliographic Procedures and Control Committee of *RLG*.

BIPS billions of instructions per second. Measure of computing power.

BIRS basic information retrieval system. Cataloguing and indexing package, Information Systems Laboratory, Michigan State University (US).

BIS 1. *Blood information service*. Database (US). 2. *Brain information service*. Database, University of California (US). 3. British imperial system. Weights and measures. 4. Bowne Information Systems. Host (US).

BISA *Bibliographic information on South-east Asia*. Database, University of Sydney (Australia).

BISAC Book Industry Systems Advisory Committee. Sub-committee of *BISG*.

BISAM basic indexed sequential access method, *IBM*.

BISG Book Industry Study Group (US).

BISITS British Iron and Steel Institute translation service.

BISNET Bank Information System network (US).

BISNY binary synchronous communication. Also *BSC*.

BISYNC bisynchronous.

BIT 1. binary digit. 2. Brighton Information Technology Conference, part of *IT82*. 3. built-in test.

BIT/s *bits* per second.

BIU basic information unit.

BIX binary information exchange.

BJF batch job foreground. Computing.

BJM between job monitor. Computing.

BJUS *Belgian jurisprudence*. Database, CREDOC.

BL 1. *BLAISE* number. Field, *OLS*. 2. block length. Computing. 3. The British Library.

BLA 1. binary logical association. 2. blocking acknowledgement signal. Telecommunications.

BLADES Bell Laboratories automatic design system (US).

BLAISE British Library Automated Information Service. Host for bibliographic databases, *BL*.

BLAISE/LOCAS see *LOCAS*.

BLBSD The British Library Bibliographic Services Division.

BLCMP Birmingham Libraries Cooperative Mechanisation Project. Now BLCMP Library Services Limited (UK).

BLEND Birmingham Loughborough electronic network development. Experimental communications network incorporating an electronic journal (UK).

BLERT block error rate test.

BLEX *Belgian lex*. Law database, *CREDOC*.

B-LINK Birmingham library and information network. Library cooperative scheme (UK).

BLIS Bell Laboratories interpretive system.

BLISS basic language for the implementation of system software.

BLLD The British Library Lending Division.

BLM basic language machine. Type of computer.

BLO blocking signal. Telecommunications.

BLOC Booth Library online circulation. Library circulation system, Eastern Illinois State University (US).

BLR&DD The British Library Research and Development Department.

BLRD The British Library Reference Division.

BLSL British Leyland Systems Ltd. Computer bureau, *IT* consultancy and development agency (UK).

BLT basic language translator, for *BASIC* language.

BLU bipolar line unit. Telecommunications.

BM binary multiply.

BMC 1. block multiplexer channel. 2. bubble memory controller. 3. bulk media conversion. 4. burst multiplexer channel. Telecommunications.

BMD 1. benchmark monitor display system, Sperry Univac (US). 2. bubble memory device.

BMG Business Machine Group, Burroughs (US).

BMI *Bibliography master index*. Database, Gale Research (US).

BMIS Bank Management Information System.

BMMC basic monthly maintenance charge.

BMMG British Microcomputer Manufacturers Group.

BMTI block mode terminal interface. Computing.

BMTT buffered magnetic tape transfer.

BN 1. binary number. 2. biosystematic code name. Searchable field, *OLS*. 3. branch name. 4. bureau number. Searchable fields, Dialog *IRS*. 5. abbreviation of *ISBN*. Searchable field, Dialog and *SDC*.

BNA Bureau of National Affairs Inc. Database originator (US).

BNB *British national bibliography*. Database, *BLBSD*.

BNC bayonette connector, for coaxial cable.

BNDO Bureau National des Données Océanique. Host specializing in oceanography, also one of its databases (France).

BNF Backus naur (normal) form. *ALGOL*.

BNF ABS *British non-ferrous abstracts*. Database, British Non-ferrous Metals Technology Centre.

BNIST Bureau National de l'Information Scientifique et Technique. Coordinating body (France).

BNPF beginning negative positive finish. Subset of *ASCII*.

BNT *Borean northern titles*. Geographical database, University of Alberta (Canada).

BNZS bipolar with N zeros substitution. *PCM* terminology.

BO binary to octel.

BOC block oriented computer.

BOCOL basic operating consumer oriented language.

BOFA *Bibliography of agriculture*. Database, *NAL*.

BOI beginning of information. Computing.

BOLD bibliographic online library display, *SDC*.

BOM Bureau of Mines. Databank originator (US).

BOOG British Osborne Owners Group. User group.

BOOGIE *BOOG* information exchange.

BOP 1. *Balance of payments*. Databank, *IMF*. 2. basic operating program. 3. binary output program. 4. bit oriented protocol. Computing.

BOR bus out register. Computing.

BORAM block oriented random access memory.

BORIS book order register and invoicing system. Microcomputer based bookshop system (UK).

BOS 1. background operating system. 2. basic operating system. 3. batch operating system. Computing. 4. book order system. Automated system, University of Massachusetts (US). 5. Business Operating Software. Software company.

BOSS 1. basic operating system, Toshiba (Japan). 2. basic operating system software, Basic Four (US). 3. batch operating software system. 4. *BLCMP* online support services. Online catalogue editing system, *BLCMP*. 5. Building Research Establishment online search system (UK). 6. business oriented software system, *DEC* (US).

BOT beginning of tape. Computing.

BP 1. batch processing. 2. bit processor. 3. buffered printing.

BPA *Bonnerville power administration*. Originator and database on power transmission (US).

BPAM basic partitioned access method. *IBM*.

BPC 1. basic peripheral channel. 2. binding post chamber. Telecommunications.

BPCC British Printing and Communication Corporation.

BPDA bibliographic pattern discovery algorithm.

BPF band pass filter.

BPI bits per inch. Storage density, computing.

BPIF British Printing Industries Federation.

BPL 1. binary program loader. 2. Burroughs program loader, Burroughs (US).

BPM 1. balanced processing monitor, Mitsubishi (Japan). 2. batch processing monitor, Xerox.

BPMM bits per millimetre. Storage density.

BPS 1. basic programming support, *IBM*. 2. *BASIC* programming system. 3. batch processing system. 4. bits per second. Transmission rate.

BPSA Business Products Standards Association (US).

BPSI bits per square inch.

BPSK binary phase shift keyed. Telecommunications.

BQL 1. basic query language. 2. batch query language. Programming languages.

BRASTACS Bradford Science Technology and Commercial Services. Information service (UK).

BRD binary rate divider.

BREMA British Radio Equipment Manufacturers Association.

BRG baud rate generator.

BRGM Bureau de Recherches Géologiques et Minières. Originator and operator of *BSS, BGI* and *PASCAL* (France).

BRI *Book review index*. Database, Gale Research (US).

BRIC Le Bureau de Recherche pour l'Innovation et la Convergence. Originator and operator of the *TRANSINOVE* databank (France).

BRIMARC Brighton *MARC* project. Collaborative cataloguing project (UK).

BRIT/MSI Bureau Robotique Informatique Télématique/Mercure Système Informatique (France).

BRM binary relationship model. Computing.

BROM bipolar read-only memory.

BROWSER browsing online with selective retrieval. Information retrieval searching system.

BRS Bibliographic Retrieval Services. Host for bibliographic databases (US).

BRT binary run tape.

BS 1. backspace. *ASCII* format effector. 2. backspace character. 3. binary subtract. 4. branch state. Searchable field, Dialog *IRS*. 5. British Standard. 6. *Bundesanstalt für strassenwesen*. English/French/German terminology bank on transportation, *BAST*.

BSAL block structured assembly language.

BSAM basic sequential access method. System procedure, *IBM*.

BSC 1. basic message switching centre. Telecommunications. 2. binary symmetric channel. 3. binary synchronous. Transmission mode. 4. binary synchronous communications. Transmission protocol.

BSCA binary synchronous communications adaptor.

BSCFL bisynchronous frame level, *NCR*.

BSCM bisynchronous communications macro.

BSC/SS binary synchronous communications/start stop.

BSD 1. Bibliographic Services Division. The British Library. 2. bulk storage device.

BSDC British Standard data code.

BSDP bibliographic service development program, *CLR*.

BSE Broadcast satellite experiment. Communications satellite (Japan).

BSELCH buffered selector channel.

B/S/Hz bits per second per Herz.

BSI British Standards Institution.

BSIE Banking Systems Information Exchange (US).

BSL bit serial link.

BSM basic storage module.

BSP Burroughs scientific processor, Burroughs (US).

BSR 1. blip scan ratio. 2. buffered send receive.

BSS 1. *Banque de données de sous-sol*. Databank on earth sciences, *BRGM*. 2. bulk storage system.

BST Business Systems Technology (US).

BT 1. barred trunk. Telecommunications. 2. British Telecom. Common carrier (UK). 3. busy tone. 4. block terminal. Telecommunications.

BTAM 1. basic tape access method. Computing. 2. basic telecommunications access method, *IBM*. 3. basic teleprocessing access method. 4. basic terminal access method. Computing.

BTC 1. block transfer controller. Computing. 2. Business and Technology Centre, *CDC* sponsored (UK).

BTD binary to decimal.

BTE bidirectional transceiver element. Telecommunications.

BTF bulk transfer facility.

BTG British Technology Group. Merger of *NRDC* and *NEB*.

BTI 1. Basic Timesharing Inc. Now BTI Computer Systems (US). 2. *British*

BTL *technology index*, now *CTI*. 3. British Telecom International. International division of *BT*.

BTL 1. beginning of tape level. 2. Bell Telephone Laboratories (US).

BTM 1. batch time-sharing monitor, Xerox (US). 2. Bell Telephone Manufacturing Co. *ITT* affiliate (Belgium).

BTMF block type manipulation facility.

BTP batch transfer program.

BTRL British Telecom research laboratories.

BTS 1. British Telecommunications Systems. Consortium for Systems X telephone switching system. 2. Business Telecommunications Services. Satellite communications network (Australia).

BTSS 1. basic time sharing system. 2. Braille time sharing system.

BTU 1. basic transmission unit. 2. British thermal unit.

BTX Bildschirmtext. Viewdata system (FRG).

BU bottom up.

BUDWSR Brown University display for working set references.

BUGS Brown University graphic system.

BUMP bottom up modular programming.

BUMS Biblioteksjanst Utlanings och Mediakontrol System. Charging and media control system (Sweden).

BUPCR Bath University programme of catalogue research. Now *CCR*.

BUR back up register.

BURISA British Urban and Regional Information Systems Association.

BVA British Videogram Association.

BVD Belgische Vereniging voor Documentatie. Belgian association for documentation, also *ABD*.

BVS Bibliotheksverbundsystem. Library network (FRG).

B/W black and white.

BWC buffer word counter. Computing.

BX branch exchange. Telecommunications.

BXB British crossbar. Telephone exchange.

BYMUX byte-multiplexer channel.

BYSINC binary communications: synchronous. Protocol, also *BCS*.

BW band width.

BWT 1. backward wave tube. 2. bandwidth ratio.

BYP bypass.

BZ branch zip code. Searchable field, Dialog *IRS*.

C

C 1. character. 2. computer language. High level language, Bell Laboratories. 3. coulomb. *SI* unit. 4. cycle.

CA 1. call number. Searchable field, Dialog *IRS*. 2. card alert. Searchable field, *OLS*. 3. channel adaptor. 4. *Chemical abstracts*. Database, *CAS*. 5. cited authors. Searchable field, *OLD*. 6. communications adaptor. 7. Computer Automation Inc. (US).

CAAS computer assisted acquisition system, libraries.

CAAT computer assisted audit techniques.

CAB 1. Civil Aeronautics Board. Originator of air transport databanks (US). 2. communications adaptor board. Computing.

CAB abstracts *Commonwealth Agricultural Bureaux abstracts*. Database on agricultural and related sciences, Commonwealth Agricultural Bureaux (UK).

CABLIS *Current awareness bulletin for librarians and information scientists*. Publication, *BL*.

CABS 1. computer aided batch searching. 2. computerized annotated bibliographic system. Alberta University (Canada). 3. *Current awareness in biological sciences*. Database, Pergamon.

CAC computer aided classification.

CACD computer aided circuit design.

CACM *Communications of the Association for Computing Machinery*. Journal (US).

CA CON *Chemical abstracts condensates*. Database, *CAS*.

CACS 1. computer aided communications system. 2. content addressable computer system.

CAD 1. character assemble disassemble. 2. computer access device. 3. computer aided design. 4. *Computer applications digest*.

CADA computer aided design analysis.

CADAPSO Canadian Association of Data Processing Service Organizations.

CADAR computer aided design and reliability.

CADC 1. central air data computer. 2. Computer Aided Design Centre. Division of *DoI* and name of its databank.

CAD/CAM computer aided design and manufacture.

CADD computer aided design and draughting.

CADE computer assisted data entry.

CADEP computer aided design of electronic products.

CADES computer aided development and evaluation system.

CADIC computer aided design of integrated circuits.

CADIG Coventry and District Information Group. Library cooperative (UK).

CADIN Canadian integration north. Radar system.

CADIS computer aided design of information systems.

CADMAT computer aided design, manufacture and test.

CADS computer analysis and design system.

CADSS combined analog digital systems simulator.

CAE 1. Compagnie Européenne d'Automatisme Électronique. Computer manufacturer (France). 2. computer aided education.

CAF computer aided fraud.

CAFS content addressable file store. *ICL IR* software.

CAG 1. Computer Applications Group, of Aslib. 2. Cooperative Automation Group. Library automation group (UK).

CAGE compiler and assembler by General Electric.

33

CAHT computer aids for human translation. Translation system, Carnegie-Mellon University (US).

CAI 1. computer aided instruction. 2. computer assisted instruction. Teaching methods. 3. Computer Automation Inc. (US).

CAIC computer assisted indexing and classification.

CAIN *Cataloging and indexing system*. Database on agriculture, National Agricultural Library (US).

CAIP computer assisted indexing program. *CAIC* system (UN).

CAIRS computer assisted information retrieval system.

CAIS 1. Canadian Association for Information Science. 2. Communication and Information Systems division of *MEP*.

CAISF *Chemical abstracts integrated subject file*. Database, *CAS*.

CAK 1. command access keys. 2. command acknowledge.

CAL 1. common assembler language, Interdata (US). 2. computer aided learning.
3. computer animation language.
4. computer assisted learning.
5. conversational algebraic language. High level language. 6. Cray assemble language, Cray Research (US).

CALCOMP California Computer Products, Inc.

CALDIS Calderdale Information Service. Library cooperative scheme (UK).

CALL Library of Congress call number. Searchable field, *SDC*.

CALLS *California academic libraries list of serials*. Union list of serials on microfiche, *CLASS*.

CALM 1. *Computer archive of language materials*, Stanford University (US).
2. computer assisted library mechanization.

CALR computer assisted legal retrieval.

CAM 1. cascade access method, *NCR*.
2. communications access method, Sperry Univac (US). 3. computer access matrix.
4. Computer aided manufacturing.
5. computer addressed memory.
6. content addressable memory.

CAMA centralized automatic message accounting.

CAMAC computer automated measurement and control.

CAMDAP *Computers and medieval data processing*. Newsletter, Institut d'Études Médiévales, Université de Montréal (Canada).

CAMERA 72 legal information retrieval project, Camera de Deputati (Italy).

CAMET *Computer archive of modern English texts*, University of Oslo (Norway).

CAMIS computer assisted make-up and imaging system.

CAMP 1. central access monitor program.
2. compiler for automatic machine programming. 3. Cooperative African microform project.

CAN cancel. Control character.

CANCERLINE *Cancer information online*. Database/databank consisting of *CANCERLIT, CANCERPROJ* and *CLINPROT*, National Cancer Institute (US).

CANCERLIT *Cancer literature*. *CANCERLINE* database.

CANCERPROJ *Cancer projects*. *CANCERLINE* database on research projects in progress.

C & C computer and communications, *NEP* (Japan).

CANDE command and edit language. Burroughs (US).

CANDOC Canadian documentation. Document ordering system (Canada). See *CANYOLE*.

C & W Cable and Wireless. Telecommunications company (UK).

CAN/MARC *Canadian machine-readable cataloguing*. Database, *NLC*.

CAN/SDI Canadian selective dissemination of information. Information dissemination service, *CISTI*.

CANSIM *Canadian socio-economic information management system*. Databank, I. P. Sharpe Associates (Canada).

CANTAT Canadian transatlantic. Cable system between Canada and UK.

CANTRAN cancel transmission.

CANUNET Canadian university computer network.

CANYOLE Canadian online enquiry. Host, National Research Council (Canada).

CAOCI *Commercially available organic chemicals index.* Databank, Chemical Notation Association (UK).

CAOS completely automatic operating system.

CAP 1. card assembly program. 2. cataloguing in advance of publication, *BLBSD*. 3. computer aided production. 4. computer aided programming. 5. Computer Analysts and Programmers Limited. Management and software consulting company (UK).

CAPDAC computer aided piping design and construction.

CAPE communications automatic processing equipment.

CAPITAL computer assisted placing in the areas of London, Employment Services Agency, Department of Employment (UK).

CAPP content addressable parallel processor.

CAPR catalog of programs.

CAPRI 1. card and printer remote interface. 2. *Chimie appliquée des polymères et des rayonnements.* Database, *CEA*. 3. computer aided personal reference index system. UK Atomic Energy Authority.

CAPS 1. computer aided problem solving. 2. *Current advances in plant science.* Database, Pergamon (UK).

CAR 1. channel address register. 2. computer assisted retrieval.

CARD compact automatic retrieval device. Massachusetts Institute of Technology (US).

CARESS career retrieval search system. Pittsburgh University (US).

CARIS computerized agricultural research information system. Information system, UN Food and Agriculture Organization.

CARL chemical algorithm for reticulation linearization.

CAROL 1. computer oriented language, Olivetti (Italy). 2. circulation and retrieval online. Library circulation system, James Cook University (Australia).

CARP computer air released point.

CARS 1. community antenna relay service. 2. computer assisted referee selection. System to aid journal editors. 3. computer assisted reference service, University of Arizona (US).

CAS 1. Chemical Abstracts Service. Database compiler (part of *ACS*), and prefix denoting some of its databases. 2. computer aid system, *SDC*. 3. computer acquisition system. 4. computerized acquisition system, R. R. Bowker Company (US). 5. control automation system, *IBM*. 6. current awareness service.

CASA Computer and Automated Systems Association (US).

CASD computer aided system design.

CASDAC computer aided ship design and construction.

CAS DDS *CAS document delivery service, ACS.*

CASE 1. computer aided software engineering. 2. computer aided system evaluation. 3. Computers and Systems Engineering. *IT* manufacturer (UK).

CAsearch *Chemical abstracts search.* Database on chemical science, *ACS*.

CAS files Chemical Abstracts Service files. Includes *CAsearch, CASIA* and *CIN*.

CASH computer assisted subject headings program. Cataloging project, University of California at San Diego (US).

CASIA *Chemical abstracts subject index alert.* Database on chemical science, *ACS*.

CASLIS Canadian Association of Special Libraries and Information Services.

CASNG *Casing design program.* Databank, Lone Star Steel Co. (US).

CASSI *Chemical abstracts source index.* Database, *ACS*.

CASSM context addressed segment sequential memory.

CAST 1. Chemical Abstract searching terminal, Computer Corporation of America. 2. computerized automatic system tester.

CASTI centers for the analysis of science and technical information (US).

CASW Council for the Advancement of Science Writing.

CAT 1. capacity activated transducer. Electronics component. 2. computer aided (assisted) teaching (training). 3. computer aided (assisted) translation. 4. computer aided typesetting. 5. computer average transients. 6. computer axial tomography. Aid to medical diagnosis.

CATCALL completely automated technique for cataloguing and acquisition of literature for libraries.

CATCH computer analysis of thermo chemical data. University of Sussex (UK).

CATED Centre d'Assistance Technique et de Documentation. Host specializing in building and construction industry (France).

CATLINE catalog online. National Library of Medicine (US).

CATNIP computer assisted technique for numerical index preparation.

CATS 1. Centre for Advanced Television Studies (UK). 2. computer aided teaching system.

CATSS catalogue support system. Developed by *UTLAS*.

CATV 1. cable television. 2. community antenna television.

CAU 1. *CPU* access unit. 2. command/arithmetic unit. 3. controller adaptor unit.

CAV constant angular velocity. Videodisc operation.

CAVE computer augmented video education, US Naval Academy.

CAW 1. Cable and Wireless. Telecommunications (UK). 2. channel address word. Data communication.

CAX community automatic exchange.

CB 1. citizens' band radio. 2. communications buffer. 3. communications bus. Computing.

CBAC *Chemical-biological activities*. Database, *CAS*.

C-BASIC commercial *BASIC*. High level language.

CBC chain block character.

CBCC CODE 22 Chemical-Biological Coordination Center code 22. Used for searching by means of chemical structure.

CBCT customer banking communication terminal.

CBDB *Conference Board data base*. Economics database (US).

CBE 1. computer based education. 2. Council of Biology Editors (US).

CBEMA Computer and Business Equipment Manufacturers Association (US).

CBI 1. Charles Babbage Institute. 2. computer based instruction.

CBL computer based learning.

CBM Commodore Business Machines. Computer manufacturer (US).

CBMA Computer and Business Equipment Manufacturers Association (US).

CBMS computer based message service/system.

CBOSS count back order and sample select. Computing.

CBPI *Canadian business periodicals index*. Database on Canadian business, industry, finance etc, Miromedia (Canada).

CBS Columbia Broadcasting System (US).

CBT computer based training.

CBX 1. centralized branch exchange. 2. Centrex service. 3. computerized branch exchange. Telecommunications.

CC 1. call check. 2. carriage control. 3. category code. Searchable field, Dialog *IRS*. 4. category code name. Searchable field, *SDC*. 5. central computer. 6. channel coordinator. 7. class code. Searchable field, Dialog *IRS*. 8. classification code. 9. colon classification. Library classification scheme. 10. communication control. 11. computer conferencing. 12. concept code. Searchable field, Dialog *IRS*. 13. condition code.

14. contractor company code. 15. country code. Searchable fields, Dialog *IRS*. 16. cursor control.

CCA 1. Central Computer Agency, Civil Service Department (UK). 2. channel to channel adaptor. 3. common communications adaptor. 4. *Computer and control abstracts*. Database, Inspec. 5. Computer Corporation of America.

CCAM conversational communication access method.

CCAP communications control application program.

CCB 1. character control block. 2. communications control block. Computing.

CCC 1. Canadian Computer Complex. 2. *Canadian criminal cases*. Databank, Canadian Law Book Limited. 3. central communications controller. 4. central computational computer. 5. central computer complex. 6. computer communication console. 7. computer control complex. 8. Copyright Clearance Center (US).

CCCC Computerized Conferencing and Communications Center, *NJIT*.

CCD 1. charge coupled device. Computing. 2. computer controlled display.

CCE communication control equipment.

CCETT Centre Commun d'Études de Télévision et de Télécommunications. Research centre active in development of videotex (France).

CCF 1. common communications format, *ISO*. 2. communications control field.

CCG Computer Communications Group, Trans-Canada Telephone System.

CCH channel check handler, Fujitsu (Japan).

CCI 1. Centro di Calcolo Interfacolta. Host, University of Rome (Italy). 2. co-channel interface. 3. Computer Communications Inc. (US). 4. Computer Composition International (US).

CCIA Computer and Communications Industry Association (US).

CCILS Centre de Calcul Interuniversitaire de Lyon Saint-Etienne. Host (France).

CCIR Comité Consultatif International des Radiocommunications. Radiocommunications standards body.

CCIS common channel interoffice signalling.

CCITT Comité Consultatif International Télégraphique et Téléphonique. Telephone and telegraph standards body.

CCL 1. common command language. 2. common control language. Computer language for searching across several computers, eg in network. 3. communications control language, *GRI* Computers. 4. *CYBER* control language, *CDC*.

CCLN Council for Computerized Library Networks (US).

CCM 1. call count meter. Telecommunications. 2. communications control module. 3. communication controller multichannel. Data communication. 4. computer coupled machines. Used in numerical control.

CCMB completion of calls meeting busy. Telecommunications.

CCP 1. channel control processor. 2. communications control package. 3. communications control processor. 4. communications control program, Sperry Univac (US). 5. console control package. 6. cross connection point. Junction in *LLN*.

CCR 1. Centre for Catalogue Research, Bath University (UK). 2. channel command register. 3. condition code register.

CCRH Center for Computer Research in Humanities, University of Colorado at Boulder (US).

CCS 1. C (100) call seconds. Telecommunications. 2. Canadian Computer Show. 3. central computer station. 4. common channel signalling. Telecommunications. 5. communications control system. 6. conversational compiling system, Xerox. 7. *CYBER* credit system, *CDC*. Combined application program and operating system.

CCSA 1. common carrier special application. 2. common channel signalling arrangement. 3. common control switching arrangement.

CCSC Coordinating Committee for Satellite Communication (Switzerland).

CCSS common channel signalling system.

CCTA Central Computer and Telecommunications Agency. Government advisory body (UK).

CCTV closed circuit television.

CCU 1. central control unit. 2. communications control unit. 3. common control unit.

CCVS *COBOL* compiler validation system.

CCW channel command/control word. Data communications.

CD 1. card. 2 carrier detect. Telecommunications. 3. check digit. 4. compact disc. Small video disc, Philips. 5. *Computer design*. Journal.

CDA 1. command and data acquisition. 2. Computer Dealers Association (US). 3. Copper Development Association. Originator and its databank (US).

CDB 1. *Coal data base*. Database, *IEA*. 2. common data bus. Computing.

CDC 1. call directing code. Telecommunications. 2. code directing character. 3. Communications and Data Centre, *BLSL*. 4. computer display channel. 5. Control Data Corporation (US). 6. *Copper data center*. Databank, *CDC*. *Cryogenic data center*. Databank, *NBS*.

CDCT Centro de Documentacao Cientifica e Técnica. Broker offering access to major hosts (Portugal).

CDCVR code converter.

CDES computer data entry system.

CDF combined distribution frame. Telecommunications.

CDI 1. collector diffused isolation. Semiconductor manufacture technique. 2. *Comprehensive dissertation index*. Database of doctoral dissertations, *UMI*. 3. Computer Devices Inc. (US). 4. Control Data Institute (US).

CDIB collector, diffusion, isolation, bipolar.

CDIS *Community data information system*. Databank, MJK Associates (US).

CDIU-PA Centre de Documentation des Industries Utilisatrices de Produits Agricoles. Database originator and operator (France).

CDK channel data check.

CDL 1. common display logic. 2. computer design language. 3. Computer Development Laboratory, Fujitsu, Hitachi and Mitsubishi (Japan).

CDMA 1. cartridge direct memory access. 2. code division multiple access. Form of multiplexing.

CDMS *COMRADE* data management system.

CDO community dial office. Small switching system.

CDP 1. central data processor. 2. Certificate in Data Processing.

CDPC commercial data processing centres.

CDR call data recording. Telecommunications.

CDS 1. Cataloging Distribution Service. Division of *LC* originating *MARC*. 2. compatible duplex system. Telecommunications. 3. *Computerized documentation system*. Database, *UNESCO*.

C/DS cache/disk system. Storage device, Univac Company.

CDSF *COMRADE* data storage facility.

CDSH *Centre de documentation sciences humaines*. Database on social sciences/humanities, *CNRS*.

CDSS computer digital switching system, Plessey.

CDST Centre de Documentation Scientifique et Technique du *CNRS*. Database originator (France).

CDT control data terminal.

CDTL common data translation language.

CDU 1. cartridge disc unit. 2. colour developing unit.

CDV check digit verification. Computing.

CDW computer data word.

CE 1. channel end. 2. customer engineer.

CEA 1. *Chemical engineering abstracts*. Database, Royal Society of Chemistry (UK). 2. Commissariat à l'Énergie Atomique. Originator and operator of databanks (France).

CEC 1. Commission of the European Communities. See *ARTEMIS, ECHO, ESPRIT, EURONET, DIANE*. **2.** Council for Exceptional Children. Produces *ECER* database (US).

CED 1. capacitance electronic disc. Videodisc. **2.** Centro Elettronico di Documentazio (Giuridica della Corte Suprema di Cassazione). Host providing information on Italian law (Italy).

CEDIJ Centre d'Information Juridique. Originator, operator and legal databank (France).

CEDOCAR Centre de Documentation de l'Armement. Host, originator and database (France).

CEF cable entrance facility.

CEGB Central Electricity Generating Board (UK).

CEH *Chemical economics handbook*, Stanford Research Institute. Online by *SDC*.

CEI computer extended instruction.

CELEX *Community lex* (ie law). Databank, *EEC*.

CEM computer education for management.

CEN 1. Centre d'Étude de l'Énergie Nucléaire. Host (Belgium). **2.** century. Searchable field, *SDC*.

CEO comprehensive electronic office. Software, Data General.

CEPII Centre d'Études Prospectives et d'Information Internationales. Databank originator (France).

CEPT Conference of European Postal and Telegraph Administrations. Telecommunications standards advisory body.

CERT character error rate test.

CERVED Centri Elettronici Reteconnessi Valutazione Elaborazione Dati (Electronic Value-added Data Network Centre). Host and database originator (Italy).

CESARS chemical evaluation search and retrieval system, *CIS*.

CESP Centre d'Étude des Supports Publicitaire. Originates and operates *MEDIA/P* (France).

CESSIM Centre d'Études des Supports d'Information Médicale. Originates and operates *MEDIA/M* (France).

CET Council for Educational Technology (UK).

CETA Chinese-English Translation Assistance Group. *MAT* system (UK).

CETEDOC Centre de Traitement Électronique de Documents. Producer of computer generated concordances to medieval Latin texts, Catholic University of Louvain (Belgium).

CETIM *Centre technique des industries Mécaniques*. Database on mechanical engineering, Centre de Documentation de la Mécanique (France).

CEU 1. channel extension unit. **2.** communications expansion unit.

CF 1. central file. **2.** *Computer fraud*. Newsletter, Elsevier. **3.** cosati field. Searchable field, Dialog *IRS*.

CFA computer family architecture.

CFB cipher feedback.

CFC 1. channel flow control. **2.** coin and fee checking. Telecommunications.

CFCE Centre Francaise du Commerce Extérieur. Databank originator.

CFCE-CPOE *CFCE – catalogue des produits francais offerts à l'exportation*. Databank of exports, *CFCE*.

CFCE-OFCE *CFCE – operators francais du commerce extérieur à l'exportation et à l'importation*. Databank, *CFCE*.

CFDE call failure detection equipment. Telecommunications.

CFES *Center for energy systems*. Databank, *GEISCO*.

CFL call failure signal. Telecommunications.

CFM 1. cathode-follower mixer. Computer graphics. **2.** cubic feet per minute.

CFP Compagnie Francaise des Pétroles. Coordinates *STATSID* databank (France).

CFSK coherent frequency shift keying. Telecommunications.

CFSTI Clearing house for Federal Scientific and Technical Information (US).

CFT Cray Fortran. High level language, Cray Research (US).

CFU current file user. Computing.

CG 1. computer graphics. 2. contract/grant numbers. Searchable field, *SDC*. 3. contributions greater than. Searchable field, Dialog *IRS*.

CGI computer generated image.

CGIS *COMSAT* general integrated system.

CGL computer generated letter.

CGP 1. central graphics processor, Evans and Sutherland. 2. colour graphics printer.

CGPC cellular general purpose computer.

CGRAM clock generated *RAM*.

CGS circuit group congestion signal, telecommunications.

CH clearing house code. Searchable field, Dialog *IRS*.

CHAIS *Consumer hazards analytical information service*, Laboratory of the Government Chemist (UK).

CHAMP 1. character manipulation procedures. 2. communications handler for automatic multiple programs.

CHAN clearing house accession number. Searchable field, *SDC*.

CHAR character.

CHARIBDIS Chalk River bibliographic data information system, Atomic Energy of Canada Limited.

CHB Compagnie Honeywell Bull. Computer manufacturer, also *CIIHB* (France).

CHCU channel control unit.

CHDB compatible high density bi-polar. Telecommunications.

CHDL computer hardware definition language.

Chemdex *Chemical index*. Database, *SDC*.

CHEMLINE *Chemical information (dictionary) online*. Database including dictionary of chemical substances, National Library of Medicine (US).

CHEMNAME *Chemical name*. Database on chemical substance names, CAS.

CHEMSDI *Chemical abstracts SDI*. Current awareness service, *CAS*.

CHEOPS Chemical Information Systems Operators, now *EUSIDIC*.

ChESS Clearing house for Education and Social Studies/Social Science, *ERIC*.

CHI computer-human interaction.

CHIF channel interface.

CHIPS clearing house interbank payment system (US).

CHKPT checkpoint.

CHOL common high order language.

CHP channel processor.

CHPS characters per second. Transmission rate.

CHRGN character generator, *ROM*.

CHS characters.

CHT call holding time. Telecommunications.

CI 1. call indicator. 2. chain index. 3. city. Searchable field, *SDC*. 4. communications interface. 5. cumulative index.

CIA 1. communications interrupt analysis, Sperry Univac (US). 2. Computer Industry Association (US). 3. computer interface adaptor.

CIB 1. channel interface bus, Honeywell (US). 2. command input buffer. Computing.

CIBD Centre d'Information des Banques de Données. Database centre, Paris (France).

CIC communications intelligence channel.

CICA Construction Industry Computing Association (UK).

CICIN Conference on interlibrary communication and information networks, 1970, *ALA*.

CICIREPATO Committee for International Cooperation in Information Retrieval among Examining Patent Offices. Also called *ICIREPAT*.

CICP communication interrupt control program.

CICRIS Cooperative Industrial and Commercial Reference and Information Service. Now West London Commercial and Technical Library Service (UK).

CICS customer information control system. Information control software used in banking (US).

CICS/VS customer information control system/virtual storage, *IBM* (US).

CID communication identifier.

CIDA channel indirect data addressing.

CIDB Chemie Information und Dokumentation, Berlin.

CIDC Consorci d'Informacio i Documentacio de Catalunya. Agency cooperating in *ESA* and *ESA-NET* (Spain).

CIDS *Chemical information and data system*. Databank, US Army management information system.

CIDSS Comité International pour l'Information et la Documentation des Sciences Sociales. Also *ICSSD* and ICSSDI.

CIDST Committee on Information and Documentation for Science and Technology, *CEC*.

CIF central information file.

CIFT *Contextual indexing and faceted taxonomic system*. Bibliographic database, *MLA*.

CIG Computer Investment Group Inc. (US).

CIGL Centre d'Informatique Général de Liège. Host and computer bureau (Belgium).

CII Compagnie Internationale pour l'Informatique. See *CIIHB*.

CIIHB Compagnie Internationale pour l'Informatique Honeywell-Bull. Computer manufacturer (France). See also *CHB, CII*.

CIJE *Current index to journals in education*. File in *ERIC* database.

CIL 1. call identification line. Message identification, teletext. 2. call information logging. Record of telephone call details. 3. Computer Investments Limited (UK). 4. current injection logic, computing.

CILE call information logging equipment.

CILEA Consorzio Interuniversitario Lombardo per l'Elborazione Automatica. Host and computer/information service consultancy (Italy).

CIM 1. communication interface monitor. 2. computer input microfilm. 3. computer input multiplexer. 4. computer integrated manufacture.

CIMI Centre d'Information de la Maison de l'Immobilier. Databank originator (France).

CIMI Drugfile Chemical Information Management Inc. *Drugfile*. Databank.

CIN *Chemical industry notes*. Database on business aspects of chemical processing industry (US).

CINDA 1. *Computer index for neutron data*. Databank, *CEA* and *IAEA*. 2. *Computer index of neutron data*. Databank, UK Atomic Energy Authority. 3. *Computer index to automated data*. Databank, Centre de Compilation de Données Neutroniques (France).

CINDAS Center for Information and Numerical Data Analysis and Synthesis. Originator and its chemical databank (US).

CINECA Consorzio Interuniversitario Nord-est di Calcolo Automatico. Cooperates in operation of Crystallographic Data Centre databases (Italy).

CINIME Centro de Informacion de Medicamentos. Originates and operates *ESPES*.

CIOCS communications input/output control system.

CIOM communications input/output multiplexer.

CIOP communications input/output processor.

CIOU custom input/output unit.

CIP 1. cataloguing in publication. 2. commercial instruction processor, Honeywell (US). 3. communications interrupt program.

CIPS Canadian Information Processing Society.

CIR 1. carrier to interference ratio, telecommunications. 2. current instruction register, computing.

41

CIRBCA *Circolari ministerio beni culturali e ambientali.* Databank on public administration, *CSC*.

CIRC circulation system. Colorado Instruments Inc. (US).

CIRC (Dundee) Ltd. Formerly Centre for Industrial Research and Consulting.

CIRCA computerized information retrieval and current awareness.

CIRK computing information technology center information retrieval from keyboards, Union Carbide Corporation (US).

CIS 1. Centre d'Informations Spectroscopiques. Information broker (France). 2. *Centre d'informations spectroscopiques.* Databank, *GAMS* and *CIS*. 3. character instruction set. 4. *Chemical information system.* Databank, *ISC, NHLI* and *EPA*. 5. commercial instruction set. 6. Compuserve Information Service. Microcomputer-based network (US). 7. Congressional Information Service Inc. Originator and databank (US). 8. Cornwall Information Service. Library cooperative (UK).

CIS abstracts *Centre International d'Informations de Securité et d'Hygiène du Travail abstracts.* Database on industrial safety and health, International Labour Office.

CISAM compressed index sequential access method.

CISCO compass integrated system compiler.

CISDC cancer information service for developing countries. Cooperative project, BL and National Cancer Institute (US), ended 1981.

CISI Compagnie Internationale de Services en Informatique. Host, data processing and computer company (France).

CISI-ELECNUL *CISI – electrical and nuclear energy.* Databank, *CISI*.

CISI-IAI *CISI – industrial activity indicators.* Databank covering OECD countries, *CISI*.

CISI-MEDIAP *CISI – media popularité.* Databank covering audience and readership statistics, *CISI*.

CISI-PI *CISI – principe indicateur.* Databank of economic indicators for the OECD, *CISI*.

CISR Center for Information Systems Research, *MIT*.

CISRC Computer and Information Science Research Center, Ohio State University (US).

CISTI Canada Institute for Scientific and Technical Information. Host and database originator, Canadian National Research Council.

CIT Compagnie Industrielle de Télécommunication (France).

CITAB computer instruction and training assistance for the blind.

CITCA Committee of Inquiry into Technological Change in Australia.

CITE 1. *Consolidated index of translations into English.* Database, *NTC* (US). 2. current information transfer in English. Interface to assist searching of databases.

CITERE Centre d'Information Temps Réel Europe. Host specializing in law, finance and technology (France).

CITM charge injection transistor memory.

CIU 1. channel interface unit.
2. communications interface unit.
3. computer interface unit. Computing.

CJ *Computer journal* (UK).

CJAIN Criminal Justice Archive and Information Network (US).

CJIS *Canadian journal of information science.*

CJPI *Criminal justice periodical index.* Database, *UMI*.

CL 1. classification code. 2. classification group. Searchable field, Dialog *IRS*. 3. clear. 4. command language. 5. computational linguistics. 6. conference location. 7. contributions less than. Searchable fields, Dialog *IRS*. 8. control language, *ICL*. 9. current loop: interface standard. 10. derwent classes. Searchable field, *SDC*. 11. derwent classification. Searchable field, Pergamon Infoline. 12. patent classification number. Searchable field, Dialog *IRS*.

CLA 1. communications line adaptor.
2. Computer Law Association (US).
3. Computer Lessors Association (US).
4. Copyright Licensing Agency (UK).

CLAIM Centre for Library and Information Management, Loughborough University (UK).

CLAIMS *Class code, assignee, index method, search*. Set of databank and databases, *IFI*/Plenum Data Co. (US).

CLAIMS/CHEM *CLAIMS chemical*. Databank of US patents and trade marks, *CLAIMS*.

CLAIMS/CLASS *CLAIMS/classification*. Databank of US patents and trademarks, *CLAIMS*.

CLAIMS/GEN *CLAIMS general*. Database on US general, electrical and mechanical patents, *CLAIMS*.

CLAIMS/US Pats Abs *CLAIMS US patent abstracts*. Databank, *CLAIMS*.

CLAS computerized library acquisitions system, Lewis and Clark College (US).

CLASS 1. California Libraries Authority for Systems and Services. Library cooperative (US). 2. centralized library automation system, Iowa State Travelling Library (US). 3. current literature alerting search service, Biological Abstracts. 4. current literature awareness service series, *ERIC*.

CLAT communications line adaptor. Buffer-teletype interface unit.

CLB central logic bus, computing.

CLC 1. central logic control, computing. 2. communications line control. 3. communications link controller, telecommunications.

CLCM Cooperating Libraries of Central Maryland. Library cooperative (US).

CLE communications line expander, telecommunications.

CLEO compiler for *LEO, ICL*.

CLF clear forward signal, telecommunications.

CLG Cooperative Libraries Group. Now Cooperative Automation Group (UK).

CLI 1. calling line identification, telecommunications. 2. command language interpreter, computing.

CLIBOC *Chinese linguistics bibliography on computer*. Cambridge University Press (UK).

CLIC 1. command language for interrogating computers, Royal Radar Establishment (UK). 2. conversational language for interactive computing.

CLINPROT *Clinical protocols*. Database on cancer clinical protocols for treatment, *CANCERLINE*.

CLIO conversational language for input/output, computing.

CLIP 1. cellular logic image processor. 2. computer language for information processing, *SDC*.

CLK clock, for synchronizing with computer system.

CLM communications line multiplexer.

CLOB core load overlay builder, General Automation (US).

CLOC collocation. Software package, University of Birmingham (UK).

CLOCK clock generator, microprocessing.

CLP communication line processor.

CLR Council on Library Resources Inc. (US).

CLS communications line switch.

CLSI Computer Library Services Inc. (US).

CLT 1. communication line terminal. 2. computer language translator.

CLU central logic unit, computing.

CLV constant linear velocity, videodisc operation.

CLW College of Librarianship Wales (UK).

CM 1. cards per minute. 2. central memory, computing. 3. community code. Searchable field, Dialog *IRS*. 4. communications multiplexer, telecommunications. 5. *Computer management*. Journal (UK). 6. control memory. 7. core memory. Computing.

CMA 1. Communications Managers Association. Trade Union, formerly *POMSA* (UK). 2. Computer Management Association (US). 3. computer monitor adaptor.

CMAP central memory access priority, computing.

CMAR control memory access register, computing.

CMC 1. code for magnetic characters.
2. communications channel.
3. communications mode control.
4. Computer Machinery Corporation (US).
5. computer mediated communications.

CMCA character mode communications adaptor.

CMD core memory driver, computing.

CME 1. central memory extension, computing.
2. concurrent machine environment, *ICL* (UK). 3. computer measurement and evaluation.

CMI 1. Cambridge Memories Inc. (US).
2. coded mark inversion. 3. computer managed instruction.

CML 1. common mode logic. 2. computer managed learning. Extension of *CAL*.
3. *COMSAT* General/*MCI* Telecommunications/Lockheed communications satellite corporation (US).
4. current mode logic, computing.

CMM 1. common memory manager, *CDC*.
2. communications multiplexer module.

CMOS complementary metal oxide semiconductor. Transistor.

CMPX *COMPENDEX*.

CMR common mode rejection, computing.

CMS 1. communications management system, Burroughs (US). 2. compiler monitoring system (computing). 3. computer management system, Burroughs (US).
4. conversational monitor system, *IBM* (US). 5. Coordinierte Management Systeme. Host and computer bureau (FRG).

CMT cassette magnetic tape.

CMU 1. Cambridge Management Unit, now *CLAIM*. 2. computer memory unit.
3. control maintenance unit.

CMX character multiplexer, telecommunications.

CN 1. citation number. Searchable field, *NLM*. 2. class name. Searchable field, Dialog *IRS*. 3. communications network.
4. company name. 5. concert name.
6. contract/grant number. 7. corporate source name. 8. country name. Searchable fields, Dialog *IRS*. 9. cross code name. Searchable field, *ESA-IRS*.

CNA 1. Chemical Notation Association (UK).
2. communications network architecture.

CNAS Chemical Nomenclature Advisory Service, Laboratory of the Government Chemist (UK).

CNC computer numerical control.

CNCP see *CN/CPT*.

CN/CPT Canadian National/Canadian Pacific Telecommunications.

CNDP communications network design program.

CNDST Centre National de Documentation Scientifique et Technique. Broker for online searching, also batch on its own computer (Belgium).

CNE 1. communications network emulator.
2. compare numeric equal.

CNET Centre National d'Études des Télécommunications (France).

CNI 1. *Canadian news index*. Database on current affairs, Micromedia (Canada).
2. changed number interception. Operator telephone service.

CNIC Centre Nationale d'Information Chimique. Database originator and host (France).

CNMI communications network management interface.

CNP 1. communications network procedure.
2. communications network processor.

CNR carrier to noise ratio, telecommunications.

CNRS Centre National de Recherche Scientifique. Major funding agency for science and originator of *PASCAL* (France).

CNRSLAB *CNRS laboratories*. Databank on research in CNRS laboratories, *CNRS*.

CNS 1. *Commodity news service*. Databank/information service on commodity prices (US). 2. communication network service. Videoconferencing. 3. Communication network service, operated by *SBS*.
4. communications network simulation.

CNT Canadian National Telephone Company.

CNU compare numeric unequal.

CNUCE Centro Nazionale Universitario di Calcalo Electronico. National University Centre for Electronic Calculation (Italy).

CO 1. coden. Searchable field, *ESA-IRS* and Dialog. 2. company. 3. contractor company name. 4. corporate source. Searchable fields, Dialog *IRS*. 5. country. Searchable field, *ESA-IRS*.

COACH Canadian Organization for the Advancement of Computers in Health.

COAM equipment customer owned and maintained communication equipment.

COBIS computer based instruction system.

COBLOS computer based loans system. Library loans system, UK Atomic Energy Authority.

COBOL common business oriented language. Computer programming language. See *CODASYL*.

COC coded optical character.

COCC character oriented communications controller.

COCO committee code. Searchable field, *SDC*.

COCOA a word count and concordance generator on *Atlas*. Atlas Computer Laboratory (UK).

COCRIL Council of City Research and Information Libraries (UK).

CODAP control data assembly program. Product of the Control Data Corporation.

CODASYL Conference on Data System Language. Agrees standards, responsible for *COBOL*.

CODATA Committee on Data for Science and Technology. Committee of the International Council of Scientific Unions.

CODE *COBOL* program development, Sperry Univac (US).

CODEC coder decoder, *PCM*.

CODES 1. computer design and education system. 2. computer design and evaluation system.

CODIC computer directed communications.

CODIL content dependent information language.

CODIS controlled digit simulator.

CODIT computer direct to telegraph.

CODOC *Cooperative documents project*, Ontario Universities Library Cooperative System. Database on official publications of commonwealth countries (Canada).

COED computer operated electronic display.

COF confusion signal, in *CCS*.

COFFI *Communications frequency and facility information systems*. Databank on aviation, *ICAO*.

COGENT compiler generator and translator, computing.

COGEODATA Committee on the Storage, Automatic Processing and Retrieval of Geological Data, International Union of Geological Sciences.

COGO coordinated geometry. High level language.

COGS concordance generation system. Text editing system, University of Toronto (Canada).

COI Central Office of Information. Government agency (UK).

COIN *COBOL* indexing and maintenance package. Software, National Computing Centre (UK).

COINS computer and information sciences.

COL 1. column. 2. communications oriented language. 3. computer oriented language.

COLA cooperation in library automation, *LASER*.

COLRS *CLASS* online reference service. Cooperative consultation and training programme.

COLT 1. communication line terminator, *IBM*. 2. computerized online testing. 3. Council on Library Technology (US).

COM 1. cassette operating monitor. 2. computer output microform/microfilm/microfiche.

COMA Computer Operations Management Association (US).

COMAC continuous multiple access collator, computing.

COMAL command algorithmic language. High level language, computing.

COMBO combination support chip, computing.

COMEINDORS composite mechanized information and documentation retrieval system.

COMET 1. *Codage de l'information métrologique.* Database originated and operated by the Bureau National de Métrologie (France). 2. computer message transmission.

COMEXT *Community external trade statistics.* Database, *EEC*.

COMM communications.

COMMANDS computer operated marketing mailing and news distribution system. Information dissemination service, Building Research Establishment (UK).

COMM MUX communications multiplexer, Perkin Elmer (US).

COMMPUTE computer oriented music materials processed for user transformation or exchange, State University of New York at Stony Brook (US).

COMMS communications interface.

COMNET 1. Communications Network. International network of documentation centres on communication research and policies. 2. Computer Network Corporation. US network, available in Europe via *TYMNET*.

COMPAC computer program for automatic control.

COMPACT 1. compatible algebraic compiler and translator. 2. *Computerization of world facts.* Databank, Stanford Research Institute (US).

COMPANDOR compressor-expandor, *FDM*.

COMPASS comprehensive assembly system, *CDC* (US).

COMPAY computer payroll.

COMPENDEX *Computerized engineering index.* Database, Engineering Index Inc. (US).

COMPOSE computerized production operating system extension. Engineering Index Inc. (US).

COMPSAC Computer Software and Applications Conference.

COMPUNICATIONS computers and communications.

COMRADE computer aided design environment.

COMSAT Communications Satellite Corporation. Technical support service (US).

COMSL communication system simulation language.

COMSTAR Internal communications satellites (US).

COMSYL communications system language.

COMTEC Computer Micrographics and Technology Group. Group of COM users and manufacturers (US).

CONCORD concordance. Concordance software, University of Münster (FRG).

CONDUIT computer based curriculum materials transportability experiment, Dartmouth College, Oregon State, Iowa and Texas Universities (US).

CONF *Conference papers index.* Database, Lockheed and *SDC*.

CONIO console input/output.

CONIT connector for networked information transfer. *IR* enhancement package, Laboratory for Information and Decision Systems, *MIT* (US).

CONLIS Committee on National Library Information Systems (US).

CONSER conversion of serials project. Database creation project (US and Canada).

CONSUL control subroutine language, computing.

CONTRAN control translator. Compiler from Honeywell.

CONTROL controller.

CONTU National Commission on New Technological Uses of Copyrighted Works (US, 1976-8).

COOKI coordinated keysort index.

COOL control oriented language.

COP 1. common online package, Fujitsu (Japan). 2. communication output printer. 3. computer optimization package, General Electric.

COPI computer oriented programmed instruction.

COPOL Council of Polytechnic Librarians (UK).

CORAL computer online real time applications language.

CORD coordinates. Searchable field, *SDC*.

CORE *Collected original resources in education*. Microfiche system, Carfax Publishing.

CORELAP computerized relationship layout planning.

CORKS computer oriented record keeping system. Central Institute for the Deaf (US).

CORS Canadian Operational Research Society.

CORSAIR computer oriented reference system for automatic information retrieval. Forsvarets Forskningsamsalt (Sweden).

CORTEX communications oriented real time executive.

COS 1. cassette operating system. 2. class of service, telecommunications. 3. communications operating system. 4. communications oriented software. 5. compatible operating system. 6. concurrent operating system, Sperry Univac. 7. core operating system, *GEC* Computers. 8. Cray operating system, Cray Research.

COSAM *COBOL* shared access method, Pertec.

COSAP cooperative online serials acquisition project.

COSATI Committee on Scientific and Technical Information. Federal committee (US).

COSBA Computer Service and Bureaux Association (UK).

COSCL common operating system control language.

COSEC Culham online single experimental console.

COSMIC Computer Software Management and Information Center, University of Georgia (US).

COSOS Conference on Self Operating Systems (US).

COSTAR conversational online storage and retrieval. Information retrieval term.

COSTI 1. Centre on Scientific and Technical Information (Israel). 2. Committee on Scientific and Technical Information (US).

COSY 1. compiler system. 2. compressed symbolic source language, *CDC*. 3. correction system.

COT 1. continuity signal, telecommunications. 2. create occurrence table. Text concordance program, University of Minnesota (US). 3. customer oriented terminal.

COTC Canadian Overseas Telecommunications Corporation.

COUPLE communications oriented user programming language.

CP 1. central processing/processor. 2. chain procedure, indexing. 3. cited pages. Searchable field, *SDC*. 4. clock pulse. 5. contract number prefix. Searchable field, Dialog *IRS*. 6. control processor. 7. control program.

CPA 1. *Chemical propulsion abstracts*. Database, Chemical Propulsion Information Agency, Johns Hopkins University (US). 2. computer performance analysis. 3. control program assist. 4. cross program auditor.

CPB 1. channel program block. 2. Corporation for Public Broadcasting (US).

CPC 1. card programmed calculator. 2. clock pulsed control. 3. channel program commands. 4. common peripheral channel. 5. computer process control. 6. computer programming concepts. 7. cycle program counter. 8. cycle program control.

CPCI 1. computer program configuration item. 2. *CPU* power calibration instrument.

CPC Library Database *Carolina Population Center library database.* Database on population dynamics of *LDCs* (US).

CPCS check processing control system, *IBM*.

CPD cards per day.

CPDAMS computer program development and management system.

CPDS computer program design specification.

CPE 1. central processing element. 2. central programmer and evaluator. 3. computer performance evaluation.

CPEBS central processing element bit slice, computing.

CPEUG Computer Performance Evaluation Users Group, National Bureau of Standards (US).

CPF control program facility.

CPFMS *COMRADE* permanent file management system.

CPFSK continuous phase frequency shift keying.

CPG 1. clock pulse generator. 2. *COBOL* program generator.

CPH characters per hour.

CPI 1. Capital Planning Information. Information consultants (UK). 2. *Central patents index*. Database, Derwent Publications (UK). 3. characters per inch. 4. computer prescribed instruction. 5. computing power index. 6. *Conference papers index*. Database, *ESA IRS* and *INKA*. 7. *Conference proceedings index*. Database, *BL*.

CPID computer program integrated document.

CPIN computer program identification number.

CPL 1. *CAST* programming language. 2. common programming language. 3. computer program library. 4. computer projects limited. 5. conversational programming language. High level language, *DEC*.

CPM 1. cards per minute. 2. *Catalogue of printed music, BL*. 3. central processor molecules. 4. computer performance management. 5. conversational program module, Fujitsu (Japan). 6. critical path method. Management technique. 7. *Current physics microform, AIP*. 8. current processor mode.

CP/M control program/microcomputers.

CPMA 1. central processor memory address. 2. Computer Peripherals Manufacturers Association (US).

CPO concurrent peripheral operations, computing.

CPOL communications procedure oriented language.

CPP 1. card punching printer. 2. *Current papers in Physics*. Titles service, *INSPEC*.

CPR card punch and reader, computing.

CP-R control program – real-time, Xerox.

CPRDC coordinated program of research in distributed computing, *SERC*.

CPS 1. card programming system, *IBM*. 2. cards per second. 3. central processor system. 4. characters per second. 5. conversational programming system. 6. *Current population survey*. Databank, *DOC*. 7. cycles per second (*Hz*).

CP-6 control program – 6. Operating system for former *CP-V* users, Honeywell.

CPSK coherent phase shift keying.

CPSM critical path scheduling method, management.

CPSP Consumer Protection and Safety Commission. Database compiler (US).

CPSS computer power support system.

CPT critical path technique, management.

CPTE computer program test and evaluation.

CPU 1. central processing unit. 2. computer processor unit.

CP-V control program – V. Operating system, Xerox.

CQMS circuit quality monitoring system.

CR 1. card reader. 2. carriage return. 3. carry register. 4. *Chemical abstracts* reference. Searchable field, Dialog *IRS*. 5. cited references. Searchable field, *ESA-IRS* and Dialog. 6. command register.

7. communications register. 8. *Computing reviews*. Journal (US). 9. contributions record. Searchable field, Dialog *IRS*. 10. control routine.

CRA catalogue recovery area.

CRAM 1. card random access memory, *NCR*. 2. core and random access manager, General Automation.

CRAR control *ROM* address register, computing.

CRAY – used with suffix – computer series, Cray Research.

CRB *Community Reference Bureau databank, EURATOM* (Italy).

CRC 1. Communications Regulatory Commission (US). 2. communication relay center, *HF*. 3. composing reducing camera, microfilming. 4. Cybernetics Research Consultants. Operators of *IR* based on *WELDASEARCH* tapes (UK). 5. cyclic redundancy check, telecommunications.

CRCC cyclic redundancy check character.

CRCGR cyclic redundancy check generator/checker, microprocessing.

CRD card reader.

CRDS *Chemical reactions documentation service*. Database, Derwent Publications (UK).

CRECORD *Congressional record*. Database (US).

CREDOC Centre de Recherche Documentaire/Centrum voor Rechtsdocumentatie. Major compiler of legal databases (Belgium).

CREL Centre des Recherches et d'Études Linguistiques (France).

CRESS computer reader enquiry service system. Automated library system.

CRFW *Catalyst resources for women*. Database, Catalyst library via *BRS*.

CRG Classification Research Group, (UK).

CRI Cray Research Inc. (US).

CRIB *Computerized resources information bank*. US Geological Survey.

CRID Centro di Riferimento Italiano Diane. Centre for *EURONET-DIANE* service (Italy).

CRIDON Centres de Recherches d'Information et de Documentation Notaires. Legal information system (France).

CRIF Centre de Recherche Scientifique et Technique de l'Industrie des Fabrications Métalliques. Host and database operator (Belgium).

CRIS current research information system. US department of Agriculture.

CRISPE *Computerised retrieval information service on precision engineering*. Database, Cranfield Institute of Technology (UK).

CRJE conversational remote job entry, computing.

CRL 1. Cellular Radio Ltd. (UK). 2. Center for Research Libraries (US).

CRM/BAM *CYBER* record manager basic access method.

CRO 1. cathode ray oscilloscope. *VDU*. 2. Copyright Receipt Office, *BL*.

CROM control read-only memory.

CROS capacitor read-only storage.

CROSS computer rearrangement of subject specialities. Biological Abstracts (US).

CROSSBOW computerized retrieval of organic structures based on Wiswesser. *IR* software for searches on chemical compounds.

CRP card reader punch.

CRQ console reply queuing.

CRS Computer Recognition Systems. *OCR* manufacturer (UK).

CRSC Center for Research on Scientific Communication, Johns Hopkins University (US).

CRT cathode ray tube.

CRTC 1. Canadian Radio, Television and Telecommunications Commission. 2. *CRT* controller.

CRTOS *CRT* operating system.

CRTU combined receiving and transmitting unit.

CRU 1. card reader unit. 2. Commodities Research Unit. Originator and its databank (UK).

CRUS Centre for Research on User Studies. Library research unit, University of Sheffield (UK).

CS 1. channel status, telecommunications. 2. chip select, computing. 3. coding specification. 4. commercial systems, Data General. 5. communications satellite, also Sakura (Japan). 6. communications services, *IBM*. 7. communications simulator, Univac. 8. communications system. 9. congress and session number. 10. contractor state code. Searchable fields, Dialog *IRS*. 11. control store. 12. conventional system, pre-coordinate indexing. 13. corporate authors. Searchable field, *ESA-IRS*. 14. corporate source. Searchable field, *ESA-IRS* and Dialog.

CSA 1. called subscriber answer, telecommunications. 2. Canadian Standards Association. 3. common services area. 4. Computer Services Association (US). 5. Computer Systems Association (US). 6. Computing Services Association (UK).

CSAM circular sequential access memory.

CSAR control store address register.

CSATA Centro Studi e Applicazioni in Tecnologie Avanzate. Originator and operator, and its databank on satellite images (Italy).

CSB called subscriber busy, telecommunications.

CSC 1. colour sub carrier, telecommunications. 2. Computer Sciences Corporation. Originates and operates financial databanks (US). 3. Computer Society of Canada. 4. Corte Suprema di Cassazione. Originator and operator of legal databanks (Italy).

CSCRF *Computer system for crop response to fertilizers*. Databank, *FAO*.

CSCSI Canadian Society for Computational Studies of Intelligence.

CSD 1. circuit switched data. 2. Civil Service Department (UK). 3. constant speed drives, computing.

CSDM continuous slope delta modulation, telecommunications.

CSDN circuit switched data network.

CSDR control store data register.

CSE 1. control systems engineering. 2. core storage element.

CSEE Canadian Society of Electrical Engineers.

CSERB Computer Systems and Electronics Requirements Board (UK).

CSG Canada Systems Group. Timesharing organization.

CSH called subscriber held, telecommunications.

CSI 1. command string interpreter, *DEC*. 2. Computer Systems International. 3. control sequence introducer, computing.

CSIR Council for Scientific and Industrial Research. Originator of *PISAL* and *SANB*, operator of scientific databases (South Africa).

CSIRO Commonwealth Scientific and Industrial Research Organisation. Major research sponsor, information broker, database originator and operator (Australia).

CSIRONET CSIRO network. Information retrieval network, *CSIRO*.

CSISAS cross section information storage and retrieval system. National Neutron Cross Section Center (US).

CSL 1. code selection language. 2. computer sensitive language. 3. computer simulated language. 4. control and simulation language. Type of *JCL*.

CSMA 1. carrier sensing multiple access. *LAN* interfacing used in Ethernet. 2. Communications Systems Management Association.

CSMA/CD carrier sensing multiple access with collision detection. Broadcast networking system.

CSO 1. Central Statistical Office. Originator and its databank on macroeconomic statistics (UK). 2. centralized service observation, network administration.

CSP control switching points, telecommunications.

C/SP communications/symbiont processor, Sperry Univac.

CSR 1. channel select register. 2. control status register.

CSS 1. character start stop. 2. command substitution system. 3. comprehensive support software, Data General. 4. Computer Sharing Services. Operator of CENCUS (US). 5. computer subsystem. 6. conversational software system, National *CSS*. 7. cordless switchboard section, telecommunications. 8. Customer Support Centre, *HP* (UK).

CSSA *Canadian social science abstracts*. Database, York University (Canada).

CSSL continuous system simulation language.

CST code segment table.

CSTPC cassette magnetic tape controller.

CSU central switching unit, telecommunications.

CSV circuit switched voice, telecommunications.

CSW channel status word, computing.

CT 1. cassette tape. 2. centre de transit. International routing term, telecommunications. 3. *Chemische technik*. Database, *DECHEMA*. 4. communications technology. 5. computer technology. 6. conference title. Searchable field, Dialog *IRS*. 7. control tag. Searchable field, *BLAISE*. 8. controlled terms. Online search vocabulary.

C/T carrier to thermal noise power ratio.

CT* control terms* major keywords. Searchable field, *ESA-IRS*.

CTA Computer Traders Association (UK).

CTB 1. code table buffer. 2. computer time bookers. 3. concentrator terminal buffer.

CTC 1. channel to channel. 2. chargeable time clock, telecommunications. 3. conditional transfer of control, computing. 4. counter/timer circuit, computing.

CTCA 1. Canadian Telecommunications Carriers Association. 2. channel to channel adaptor.

CTCP *Clinical toxicology of commercial products*. Databank, *NHL* and *EPA*.

CTD charge transfer device.

CTE computer telex exchange, *RCA*.

CTG Communications Task Group, *CODASYL*.

C13-NMR *Carbon 13 nuclear magnetic resonance*. Databank, *INKA*.

CTI 1. Centre de Traitement de l'Information. Host (Belgium). 2. *Communication technology impact*. Newsletter, Elsevier. 3. Computer Translation Inc. (US). 4. *Current technology index*. Library Association Publishing Ltd (UK).

CTL 1. cassette tape loader. 2. complementary transistor logic. 3. Computer Technology Limited (UK). 4. core transistor logic.

CTMC communication terminal module controller, Sperry Univac.

CTNE Compania Telefonica Nacional de Espana. *PTT*, originator and operator of *EXPOQUIMIA* databank (Spain).

CTOS cassette tape operating system, Datapoint.

CTP command translator and programmer.

CTS 1. cable turning system, telecommunications. 2. Canadian Technology Satellite. 3. clear to send, data communications. 4. common test subroutine, computing. 5. communications technology satellite. 6. communications terminal synchronous. 7. computer typesetting. 8. conversational terminal system. 9. conversational time sharing, computing.

CTSI Computer Transceiver Systems Inc. (US).

CTS/RTS clear to send/request to send.

CTSS compatible time sharing system, *MIT*.

CTU 1. Central Telephone and Utilities Corporation (US). 2. central terminal unit, computing.

CTUK Computer Town United Kingdom. Computer literacy project.

CU 1. control unit, computing. 2. crosstalk unit, telecommunications.

CUA computer users association.

CUBE Cooperating Users of Burroughs Equipment.

CUDN common user data network.

CUE 1. computer updating equipment. 2. control unit end.

CUG closed user group, of viewdata users.

CULP *California union list of periodicals.*

CULT Chinese university language translation. *HAMT* system (Hong Kong).

CUMARC cumulated *MARC.*

CUP 1. Cambridge University Press (UK). 2. communications user program, Sperry Univac.

CUSP commonly used system programs, Digital Equipment Corporation.

CUSS 1. computerized uniterm search system. *IR* system, Dane and More (US). 2. cooperative union serials system. Ontario University Libraries Cooperative System (Canada).

CUT control unit tester, Sperry Univac.

CUTS cassette user tape system.

CV 1. cited volume. Searchable field, *SDC*. 2. consumer video. Sony (Japan).

CVC carrier virtual circuit, telecommunications.

CVIS computerized vocational information system.

CVM *COBOL* virtual machine.

CVR computer voice response.

CVSD continuously variable slope delta modulation, telecommunications.

CVT communications vector table.

CW 1. call waiting, telecommunications. 2. command word. 3. *Computer world.* Journal (US). 4. continuous wave. 5. control word. 6. corporate word. Searchable field, *BLAISE.*

CWA 1. *Canada water.* Database, Environment Canada. 2. control word address.

CWI calls waiting indicator, telecommunications.

CWP 1. communicating word processors. 2. current word pointer.

CWTG Computer World Trade Group (UK).

CX central exchange, telecommunications.

CXA *CX* area.

CY 1. conference year. Searchable field, Dialog *IRS*. 2. country. Searchable field, *ESA-IRS.*

CYBER with suffix, computer series, *CDC.*

CYCLADES packet switching network (France).

CZE compare zone equal, computing.

CZU compare zone unequal, computing.

D

DA 1. data acquisition. 2. data administrator. 3. decimal add. 4. decimal to analog. 5. define area. 6. design automation. 7. destination address. 8. differential analyser. 9. digital to analog. 10. direct address. 11. discrete address. 12. display adaptor.

D/A digital to analog.

DAA 1. data access arrangement. Device to prevent mutual damage between telephone and high speed fax equipment. 2. direct access arrangement. Telecommunications device.

DAAS *Drilling activity analysis system.* Databank, Petroleum Information Corporation (US).

DAB 1. display arrangement bits. 2. display assignment bits. 3. display attention bits.

DAC 1. data acquisition and control. 2. demand assignment controller. 3. digital analysis converter. 4. digital to analog converter. 5. display analysis console.

DA/C data acquisition.

DACBU data acquisition control and buffer unit.

DACC direct access communication channel.

DACE data acquisition and control executive.

DACOM 1. data communication. 2. datascope computer output microfilmer. *COM* device.

DACOS data communication operating system.

DAC system data acquisition and control system. Computer system.

DACU digitizing and control unit.

DACVR digital analog converter.

DADB data analysis data base.

DADOC *Direktabfrage für dokumentation.* Database on armed forces, Landesverteidigungsakademie (Austria).

DADS data acquisition display subsystem.

DAEDAC Drug Abuse Epidemiology Data Center. Operator and originator of *ODF* databank, Texas Christian University (US).

DAF 1. destination address field. 2. distributed acquisition facility.

DAFM direct access file manager.

DAFT digital analog function table.

DAGAS dangerous goods advisory service, Laboratory of the Government Chemist (UK).

DAIR 1. driver aid information and routing, computing. 2. dynamic allocation interface routine, computing.

DAISY 1. Dairy information system. Microcomputer based system developed at Reading University (UK). 2. decision aided information system. 3. double precision automatic interpretive system. Bendix Corporation.

DAL 1. data address line. 2. direct address line.

DALK data link controller, computing.

DAM 1. data association message. 2. descriptor attribute matrix. 3. direct access memory. 4. direct access method, Sperry Univac.

DAMA demand assignment multiple access.

DAMIT data analysis Massachusetts Institute of Technology (US).

DAMP *Databank of atomic and molecular physics.* Originated and operated by Queen's University, Belfast (Northern Ireland).

DAMPS data acquisition multiprogramming system, *IBM*.

DAMS direct access management system.

DAN distribution analysis. Concordance program, University of Minnesota (US).

D and B Dunn and Bradstreet. Originator, *DMI* (US).

DAP 1. data access protocol, *DEC*. 2. digital assembly program, computing. 3. distributed array processor, Sperry Univac.

DAPS 1. direct access programming system. 2. distributed application processing system.

DAPU data acquisition and processing unit.

DAR 1. daily activity report. 2. damage assessment routine. 3. data access register. 4. destination access register.

DARE 1. *Data retrieval system for documentation in the social and human sciences*. Database, *UNESCO*. 2. *Dictionary of American regional English*. 3. documentation automated retrieval equipment.

DARES data analysis and reduction system.

DARMS digital alternate representation of musical symbols. Input code for music notation.

DAS 1. data access security. 2. data acquisition system/subsystem. 3. data analysis system. 4. digital analog simulator.

DASD direct access storage drive, computing.

DASDL data and structure definition language.

DASDR direct access storage dump restore.

DASF direct access storage facility.

DASH dual access storage handling.

DASL data access system language.

DASS digital access signalling system.

DAT 1. disc allocation table. 2. dynamic address translation.

DATAC data analog computer.

DATACOM 1. data communication. 2. data communication service, Western Union (US). 3. data communications network, US Air Force.

DATEL data telecommunication. Data transmission service (UK).

DATEV Datenverarbeitungsorganisation des Steuerberaten berufes in der Bundesrepublik Deutschland. Originator and operator of legal databank (FRG).

Datex-P data text. Data transmission network (FRG).

DATICO digital automatic tape intelligence checkout.

DATRAN Data Transmission (US).

DAU 1. data access unit. 2. data adaptor unit.

DAV data above voice system, telecommunications.

DAVID data above video system.

DB 1. data bank. 2. data base. 3. data bus. 4. decibel. 5. decimal to binary. 6. Deutsche Bundespost, *PTT* (FRG).

DBA database administrator.

DBAAM disc buffer area access method.

DBACS database administrator control system.

DBAF database access facility.

DBAM database access module/method.

DBAWG Data Base Administration Working Group, *CODASYL*.

DBCB database control block.

DBCL database command language.

DBCS database control system.

DBD database diagnostics.

DB/DC database/data communications.

DBDL database definition language.

DBG database generator.

DBI 1. *Database index*. Database, *SDC*. 2. Deutsches Bibliothekinstitut. Library institute (FRG). 3. double-byte interleaved.

DBIN data bus in.

DBIOC database input/output control.

DBLTG Data Base Language Task Group, *CODASYL*.

DBM 1. database management. 2. decibels to one milliwatt. Unit of signal strength. 3. direct branch mode.

DBMO decibels relative to one milliwatt at a point of zero relative level. Unit of signal strength.

DBMOP *DBMO* and psophometrically weighted for telephony. Unit of signal strength.

DBMOPS *DBMO* and psophometrically weighted for sound programme transmission. Unit of signal strength.

DBMS database management system.

DBMSPSM database management system problem specification model.

DBOMP database organization and maintenance processor.

DBOS disc based operating system, computing.

DBP database processor.

DBQ decibels (relative to a reference voltage defined by *CCITT*) measured with a quasi-peak noise meter (without a weighting network). Measure of absolute voltage level of audio-frequency noise.

DBR 1. database retrieval. 2. decibel relative level. Unit of noise.

DBS 1. data base supplier. 2. database system. 3. direct broadcast satellite, to local or domestic receiver.

DBTG Database Task Group, *CODASYL*.

DBU digital buffer unit.

DC 1. data cartridge. 2. Datacentralen. Host and computer bureau (Denmark). 3. data channel. 4. data conversion. 5. decimal classification. Library classification scheme and searchable field, *BLAISE*. 6. descriptor code. Searchable field, *NLM* and Dialog. 7. design change. 8. detail condition. 9. device control. 10. digital computer. 11. direct coupled. 12. direct current. 13. disc controller. 14. display code. 15. display console.

DCA 1. data communications administrator. 2. Defense Communications Agency (US). 3. distributed communications architecture, Sperry Univac. 4. document content architecture, for *WPP* and *EDD*.

DCAM data collection access method.

DCAS digital control and automation system.

DCB 1. data control block. 2. data control bus, computing.

DCC 1. data channel converter. 2. data circuit concentration. 3. data communication channel. 4. data communications controller, computing. 5. digital cross current. 6. direct computer control. 7. direct control channel.

DCCS distributed capacity computing system.

DCCU data communications control unit.

DCD 1. data carrier detect. Data communication signal. 2. dynamic computer display.

DCDC 1. data communication to disc control. 2. direct current to direct current.

DCE 1. data circuit terminating equipment, for channel interfacing. 2. data communications equipment. 3. data conversion equipment. 4. digital control element, computing.

DCF 1. disc controller formatter. 2. document composition facility.

DCI 1. Data Courier Inc. Database producer, *ABI/Inform, CPI, PNI* etc. (US). 2. direct channel interface.

DCIO direct channel interface option.

DCL digital command language *DEC*.

DCM 1. data communications multiplexer. 2. display control module.

DCMS dedicated computer message switching.

DCN distributed computer network.

DCNA data communication network architecture.

DCO digital central office, telecommunications.

DC1, DC2 etc device control characters.

DCOS data collection operating system.

DCP 1. data collection platform, weather satellite system. 2. data communications processor. 3. digital computer programming. 4. distributed communications processor.

DCPC dual channel port controller.

DCPCM differentially coherent pulse code modulation.

DCPP data communication preprocessor.

DCPSK differentially coherent phase shift keying.

DCR 1. data conversion receiver. 2. design change recommendation. 3. detail condition register. 4. digital cassette recorder. 5. digital condition register. 6. digital conversion receiver.

DCRABS disc copy restore and backup system.

DCS 1. data collection system. 2. data communications system, *IBM*. 3. data control services. 4. data control system, Burroughs. 5. Defense Communications System (US). 6. diagnostic control store. 7. digital command system. 8. direct coupled system. 9. distributed computer system.

DCSCS data code and speed conversion subsystem.

DCT data communications terminal.

DCTL direct coupled transistor logic.

DCU 1. data capture unit, computing. 2. data communications unit. 3. data control unit. 4. device control unit. 5. disc control unit.

DCW 1. data control word. 2. dynamic channel exchange.

DD 1. data definition statement. 2. data demand. 3. data dictionary. 4. deadline data. Searchable field, *SDC*. 5. decimal display. 6. decimal divide. 7. decision data. 8. delay driver. 9. Dewey decimal number. Searchable field, Dialog and *SDC*. 10. digital data. 11. digital display. 12. document delivery. 13. double density.

DDA 1. digital differential analyser. 2. digital display alarm. 3. direct data attachment. 4. direct device attachment. 5. direct disc attachment.

DDAS digital data acquisition system.

DDB 1. distributed data base. 2. *Dortmund data bank*. Chemical databank, Universität Dortmund (FRG).

DDBG *DIMDI* database generator. *ISR* software, *DIMDI*.

DDBS descriptor database system, *NEC/NTIS*.

DDC 1. Defense Documentation Center. Database and databank producer, eg *WVIS* and *TR* (US). 2. *Dewey decimal classification*. Library classification scheme. 3. digital data conversion. 4. digital display conversion. 5. direct digital control, computing.

DDCE digital data conversion equipment.

DDCMP digital data communications message protocol, *DEC*.

DDCS 1. data definition control system. 2. digital data calibration system.

DDD direct distance dialling. Telecommunications.

DDE 1. direct data entry. 2. distributed data entry.

DDES direct data entry station.

DDF database definition file, defining database file structure.

DDG 1. digital data generator. 2. digital display generator.

DDGE digital display generator element.

DDI direct dialling in. Routes telephone calls direct to extension.

DDL 1. data definition language. 2. data description language. 3. digital design language.

DDLC Data Description Language Committee, *CODASYL*.

DDLG data definition language committee.

DDM digital display makeup.

DDMA disk direct memory access.

DDN digital data networks.

DDOCE digital data output conversion equipment.

DDP 1. digital data processor. 2. distributed data processing.

DDR 1. device dependent routine. 2. digital data recorder. 3. dynamic device reconfiguration, Fujitsu (Japan).

DDS 1. data dictionary system, *ICL*. 2. data dissemination system, *IRS*. 3. Dataphone digital system. *A T & T* Network (US). 4. digital display scope.

DDT 1. data description table. 2. *DIBOL* debugging technique, *DEC*. 3. digital data transmitter. 4. digital debugging tape. 5. dynamic debugging technique.

DDTE digital data terminal equipment.

DDTU digital transfer unit.

DDX 1. Dendenkosha Digital Data Exchange, *NTT*. 2. digital data exchange.

DE 1. data entry. 2. decision element. 3. descriptor. Searchable field, *ESA-IRS* and Dialog. 4. descriptor entry version. Searchable field, *NLM*. 5. digital element. 6. display element. 7. display equipment. 8. division entry.

DEA data encryption algorithm.

DEAC data exchange auxiliary console.

DEAFNET network for the deaf. Electronic mail (US and UK).

DEAL 1. data entry application language. 2. decision evaluation and logic.

DEC Digital Equipment Corporation (US).

DECADE *DEC* automatic design.

DECB data event control block.

DECDR decoder.

DECHEMA Deutsche Gesellschaft für Chemisches Apparatwesen. Originates and operates *CT* and *AgChemDok* (FRG).

DECLAB *DEC* laboratory.

DECNET *DEC* network.

DECS data entry control system.

DECUS Digital Equipment Computer Users' Society, for *DEC* users.

DED 1. data element dictionary. 2. double error detection.

DEDS dual exchangeable disc storage.

DEE digital evaluation equipment.

DEEDS documents of Essex England data set. System for analysis of medieval charters, University of Toronto (Canada).

DEEP data exception error protection.

DEF 1. data entry facility. 2. data extension frame.

DEFT dynamic error free transmission.

DEG degree. Searchable field, *SDC*.

DEL 1. data entry language. 2. delete. 3. Delft Hydraulics Laboratory. Database originator (Netherlands). 4. direct exchange line, telecommunications.

DELTA distributed electronic test and analysis.

DEM demodulator.

DEMA Data Entry Management Association.

DEMOD demodulator.

DEMUX demultiplexer, telecommunications.

DEOT disconnect end of transmission.

DEP 1. data entry panel. 2. diagnostic executive program, *SEL*.

DER directly executable representation.

DERA *Directory of education research and researchers in Australia*. Database, *AUSINET*.

DERS data entry reporting system.

DES 1. data encryption standard, *NBS*. 2. data entry system. 3. descriptor(s). 4. design and evaluation system. 5. digital expansion system.

DESC data entry system controller.

DESI *Drug efficiency study*. Databank, *FDA*.

DESY Deutsches Elektronen Synchrotron. Originates and operates *HEP* (FRG).

DETAB decision table.

DETAP decision table processor.

DETOC decision table to *COBOL*.

DETRAN decision translator.

DEU 1. data encryption unit. 2. data entry unit. 3. data exchange unit.

DEUA Digitronics Equipment Users Association.

DEUCE digital electronic universal calculating engine.

DEVIL Direct evaluation of index languages. Study by *INSPEC*.

DEX 1. data exchange. 2. deferred execution.

DEXT distant end cross-talk, telecommunications.

DF 1. data field. 2. destination field. 3. direct flow. 4. direction finding. 5. disc file. 6. domestic or foreign recipient. Searchable field, *IRS*. 7. dual facility.

DFA distributed function architecture.

DFAST dynamic file allocation system.

DFB data flag branch, computing.

DFBM data flag branch manager. *DFB* routine.

DFBR data flag branch register. *DFB* hardware.

DFC 1. data flow control. 2. disc file check. 3. disc file controller.

DFCNV disc data file conversion program, *IBM*.

DFCU disc file control unit.

DFEU disc file electronics unit.

DFF display format facility.

DFI disc file interrogate.

DFL display formatting language.

DFN data file number.

DFO 1. directed format option. 2. disc file optimizer.

DFPL data flow programming language.

DFS distributed file system.

DFSU disc file storage unit.

DFT diagnostic function test. Software to test system's reliability.

DG 1. Data General Corporation. Computer manufacturer (US). 2. degree year. Searchable field, Diaglog. 3. differential generator. 4. Digital Group (US).

DGBC digital geoballistic computer system.

DGC Data General Corporation. Computer manufacturer (US).

DG/CS Data General/communications system.

DGD Deutsche Gesellschaft für Dokumentation. German Society for Documentation (FRG).

DG/DBMS Data General/data base management system.

DG/L Data General/programming language.

DGM Deutsche Gesellschaft für Metallkunds. Operates *SDIM* (FRG).

DGRST Délégation Général à la Recherche Scientifique et Technique. Originator, operator and its databank (France).

DGS 1. data gathering system. 2. display generating system.

DGSS distributed graphics support subroutines, Tektronix.

DGT Directorate Général de Télécommunication. *PTT* authority (France).

DG XIII Directorate-General (Section XIII) Part of *CEC* promoting scientific and technical information and communication, included in *CIDST*.

DG/TPMS Data General/transaction processing management system.

DH 1. decimal to hexadecimal. 2. device handler.

DHE data handling equipment.

DHLLP direct high level language processor.

DHP document handler processor.

DHU document handler unit.

DI 1. data input. 2. Department of Industry (UK). Also *DOI*, now *DTI*. 3. device independent. 4. digital input.

DIA 1. digital input adaptor. 2. direct interface adaptor. 3. document interchange architecture, *EDD*. 4. dual interface adaptor.

DIAD digital image analysis and display. *PXEL* based system.

DIADEM *Dynamic international access to databases and economic models.* Databank, *EML*.

DIAL 1. display interface assembly language, *DEC*. 2. drum interrogation alteration and loading. 3. structure and representation of data for interchange at the application level. Recommendations for computer to computer communication.

Dialorder document delivery system, not an acronym, Lockheed.

DIAM data independent architecture model.

DIAN digital analog simulator.

DIANE direct information access network for Europe. *IRS* operated by *CEC* via *EURONET*.

DIAS *DIMDI*'s administration system. *DIMDI* software.

DIB 1. *Daily intelligence bulletin*. Greater London Council Intelligence Unit Research Library (UK). 2. Data Inspection Board. Data protection enforcement agency (Sweden).

DIBITS di-binary digits. Groups of two bits used in signalling.

DIBOL digital business oriented language, *DEC*.

DIBS digital integrated business system, *DEC*.

DIC 1. digital input control. 2. digital interchange code.

DICAM datasystem interactive communications access method, *DEC*.

DICIS Duane Information Center indexing service. Database compilers (US).

DID 1. data item description. 2. *Datamation industry directory*. 3. device identifier. 4. digital information detection. 5. digital information display. 6. direct in(ward) dialling. Call routing direct to extension (US).

DIDAD digital data display.

DIDAP digital data processor.

DIDM document identification and description macros, *IBM*.

DIDO 1. data input/data output. 2. digital input/digital output.

DIDS domestic information display system. Computer graphics system (US).

DIEN data input ensemble.

DIF device input format.

DIFU Deutsches Institut für Urbanistik. Originator and operator of databases on engineering and public administration (FRG).

DIG digital input gate.

DIGICOM digital communications system.

DIL dual inline, electronic components.

DILIC dual inline pinned integrated circuit, electronics.

DILS Dataskil integrated library system. Software, *ICL* (UK).

DIM 1. data and instruction management machine, *CIIHB*. 2. data interpretation module. 3. device interface module.

DIMDI Deutsches Institut für Medizinische Dokumentation und Information. Host and abstracting and indexing service in medicine (FRG).

DIMECO *Dual independent map encoding file of countries.* Databank of boundaries expressed as coordinates, Harvard University (US).

DIMS distributed intelligence microcomputer system.

DIN 1. Deutsche Industrie Normen. Standards authority (FRG). 2. Deutsches Institut für Normung. Originates and operates *NIB* databank.

DINA distributed information processing network architecture, *NEC/NTIS*.

DINUPS *DIMDI*'s input and updata system. *DIMDI* software.

DIO 1. data input/output. 2. direct input/output.

DIOB digital input/output buffer.

DIOC digital input/output control.

DIODE digital input/output display system. Element in computer system.

DIOP double-density disk drive input/output processor, Cray Research.

DIOS 1. direct memory access input/output system, Perkin-Elmer. 2. distributed input/output system. 3. distributed input/output system.

DIP 1. display information processor. 2. distributed information processing. 3. *Dokumentations und informationssystem für parllamentsmaterial.* Database on legislation and parliamentary debates, Bundestag (FRG). 4. dual inline package, for *ICS.* 5. dual inline pin. See *DIL.*

DIPS development information processing system.

DIR 1. data input register. 2. discipline oriented information retrieval. 3. document information retrieval.

DIRECT document information retrieval and evaluation for the computer terminal. Philips (Netherlands).

DIRS *DIMDI* information retrieval system.

DIRS 3 *DIMDI*'s information retrieval system 3.

DIS 1. distributed information system. 2. draft international standard. Provisional *ISO* standard.

DISAC digital simulator and computer.

DISAM direct and index sequential access system, *BLCMP.*

DIS RX disable receive. Status activation code.

DISS 1. digital interface switching system. 2. distributed information processing service system, *NEC/NTIS.*

DISSPLA display integrated software system and plotting language.

DISTL Display Electronics communications. UK microcomputer network, Display Electronics.

DIS TX disable transmit. Status activation code.

DITRAN diagnostic *FORTRAN.*

DIU 1. digital input unit. 2. digital interchange utility, *IBM.*

DIV data in voice system, telecommunications.

DIVA digital input voice answerback.

DIVOT digital to voice translator.

DJNR Dow Jones news retrieval. Information retrieval service on finance (US).

DJSU digital junction switching unit, telecommunications.

DKFZ Deutsches Krebsforschungszentrum. Operates *CANCERNET* (FRG).

DKI 1. data key idle. 2. Deutsches Krankenhausinstitut. Originates and operates *LIT-KRAN* (FRG). 3. Deutsches Kuntstoff Institut. Originator and operator of *KKF* (FRG).

DL 1. data link. 2. diode logic. 3. dual language.

DLA 1. data link adaptor. 2. Division of Library Automation, University of California, formerly *ULAP* (US).

DLC 1. data link control. 2. duplex line control, telecommunications.

DLCN distributed loop computer network.

DLE data link escape. Control character.

DLL dial long lines. Class of range extender, telecommunications.

DLM data line monitor.

DLMCP distributed loop message communication protocol.

DLMF *Drug literature microfilm file.* Database, *IDIS.*

DLO *Décisions de l'Orateur.* Full text database of parliamentary speaker's rulings (Canada).

DL/1 data language 1.

DLOS distributed loop operating system.

DLP 1. data link processor, Burroughs. 2. data listing programs.

DLPG *DIMDI* list program generator. *DIMDI* software.

DLR *Dominion law reports.* Databank, Canada Book Ltd.

DLS digital line system, telecommunications.

DLT 1. data line terminal. 2. data loop transceiver. 3. decision logic table. 4. decision logic translator. 5. dual language translation, Chinese University of Hong Kong.

DLU 1. data line unit. 2. digital line (terminating) unit, telecommunications.

DM 1. data management. 2. data memory. 3. decimal multiply. 4. delay modulation. 5. delta modulation, telecommunications.

DMA direct memory access. Computing method.

DMAC direct memory access controller.

DMACP direct memory access communications processor.

DMAI direct memory access interface.

DMB data management block.

DMC 1. direct memory channel. 2. direct multiplexed channel.

DMCL device/media control language, for designating storage locations, *CODASYL/ Honeywell*.

DME 1. digital multiplex equipment, telecommunications. 2. direct machne evironment, *ICL*.

DMED digital message entry device.

DMF 1. data management facility. 2. disc management facility.

DMH device message handler, *IBM*.

DMI 1. direct memory interface. 2. *Dunn's market identifiers*. Database, *D and B*. 3. dynamic memory interface.

DML 1. data management language. 2. data manipulation language, *DEC*.

DMM 1. data manipulation mode. 2. *Defence market measures*. Database on Department of Defense contracts (US). 3. direct memory management.

DMNSC digital main network switching system, telecommunications.

DMOD delta modulation.

DMOS 1. data management operating system. 2. discrete metal oxide semiconductor. 3. double diffused metal oxide semiconductor, microelectronics.

DMP direct memory processor.

DMQ direct memory queue, computing.

DMR 1. data management routine. 2. distributed message router, data communications.

DMS 1. data management system. 2. database management system. 3. disc monitor system, *IBM*. 4. display management system, *IBM*. 5. distributed maintenance services, Honeywell. 6. dynamic mapping system, *HP*.

DMT direct memory transfer, computing.

DM2 *Defense market measures system*. Databank on defence and space contracts, *F and S*.

DMU distributed microprocessor unit.

DMUS data management utility system.

DMX 1. data multiplexer. 2. direct memory exchange.

DN 1. data name. 2. decineper. Reference unit, similar to *DB*. 3. document number.

DNA digital network architecture, *DEC*.

DNAM data network access method.

DNC direct numerical control, of machines. Automation method.

DNCC data network control centre, network administration.

DNCS distributed network control system.

DNIC data net identification code.

DNS distributed network system.

DNSC 1. data network service centre. 2. digital network service centre, network administration.

DNT digital network terminator.

DO 1. data output. 2. decimal to octal. 3. digital output. 4. donor. Searchable field, Dialog *IRS*.

DOA digital output adaptor.

DOB data output bus.

DOBIS Dortmunder online Bibliotheksystem. Integrated library system, University of Dortmund and *IBM* (FRG).

DOC 1. Department of Communications. Federal department (Canada). 2. US Department of Commerce. Databank originator. 3. digital output control. 4. document.

DOCFAX document facsimile transmission.

DOCLINE document ordering online. Document delivery system, *MEDLARS*.

DOCOCEAN *Documentation océanique*. Database on oceanography, *BNDO*.

DOCS disc oriented computer system.

DOCSYS display of chromosome statistics system.

DOCUS display oriented computer usage system.

DOD 1. digital optical disc. Storage medium. 2. direct outward dialling, telecommunications.

DODT display octal debugging technique.

DOE Department of Energy. Database producer (US).

DOES *Directory of educational software* (UK).

DOF device output format.

DoI Department of Industry, now *DTI* (UK).

DOIO directly operable input/output.

DOIT 1. database oriented interrogation technique. 2. digital output/input translator.

DOKDI Dokumentationsdienst der Schweizerischen Akademie der Medizinischen Wissenschaften. Online information broker (Switzerland).

DOL display oriented language.

DOM disc operating monitor.

DOMA Dokumentation Maschinenbau. Originator, operator and database on mechanical engineering (FRG).

DOMSAT domestic satellite. Communications satellite (Australia).

DONA decentralized open network architecture, *OKI* Electric.

D1, D2 etc. *PCM* systems series, *ATT*-Bell.

DOPS digital optical projection system. Radar plotting system.

DOR 1. data output register. 2. digital optical recording. Videodisc information recording technique. 3. digital output relay.

DORIS direct order recording and invoicing system. Shell (UK).

DORK diagnostically optimizable recursive keyword. Program generator.

DOS 1. digital operating system. 2. disc operating system, magnetic discs.

DOS/VS disc operating system/virtual storage, *IBM*.

DOS/VS-AF *DOS/VS* – advanced functions, *IBM*.

DOS/VSE *DOS/VS* extended, *IBM*.

DOT 1. digital optical technology system. 3D *TV* system. 2. *Direction of trade*. Databank, *IMF*.

DOTIC *Directory of title pages indexes and contents pages*, UK Serials Group.

DOTIR *Dottrina e dibattiro (giuridico)*. Legal database (Italy).

DOTSYS dot system. Braille translation system, Mitre Corporation.

DP 1. data pointer. 2. data processing. 3. date of publication. *OLS* field. 4. digital plotter. 5. disc pack. 6. distribution point, in *LLN* or *CATV*. 7. drum processor. 8. dynamic programming.

DPA 1. Data Protection Act (UK). 2. data protection agency.

DPC 1. data processing centre. 2. data processing control. 3. direct program control. 4. display processor code.

DPCM 1. differential pulse code modulation. Transmission technique. 2. distributed processing communications module.

DPCTG Database Program Conversion Task Group, *CODASYL*.

DPCX distributed processing control executive.

DPD digital plane driver.

DPDL distributed program design language.

DPE 1. data processing equipment. 2. direct plate exposer, printing. 3. distributed processing environment.

DPEX distributed processing executive program.

DPF dual program feature.

DPG digital patter generator.

DPI data processing installation.

DPL data processing language.

DPLL digital phase lock loop.

DPM 1. data processing machine. 2. data processing manager. 3. documents per minute.

DPMA Data Processing Management Association (US).

DPMC dual port memory control.

DPMOAP Society of Data Processing Machine Operators and Programmers (US).

DPO data processing operations.

DPP digital parallel processor.

DPPX distributed processing programming executive, *IBM*.

DPS 1. data processing system. 2. disk programming system, *IBM*. 3. distributed present services, *IBM*. 4. distributed processing system, Honeywell. 5. document processing system, *IBM*. 6. dynamic processing system, Mitsubishi.

DPSA Data Processing Supplies Association (US).

DPSK differential phase shift keying, telecommunications.

DPSS data processing system simulator.

DPTX distributed processing terminal exchange, Prime Computers.

DPU 1. data processing unit. 2. digital patch unit. 3. disc pack unit.

DPWM double-sided pulse-width modulation, telecommunications.

DQ directory enquiry, telecommunications service.

DR 1. data recorder. 2. data reduction. 3. data register.

DRA data resource administrator.

DRAFT document read and format translator.

DRAM dynamic *RAM*.

DRAW direct read after write. Technique for correction of information on optical discs.

DRC 1. data recording control. 2. data reduction compiler.

DRCS dynamically redefinable character set. Videotex display technology.

DRD data recording device.

DRE *Dokumentationsring electrotechnik*. Database, *ZDE*.

DREAM data retrieval entry and management.

DREF *Data reference system*. Database on environmental science and issues, Environment Canada.

DRI 1. Data Recording Instruments Ltd. Peripheral manufacturer, subsidiary of *DD* (UK). 2. data reduction interpreter. 3. Data Resources Inc. Database originator and operator (US). 4. document retrieval index.

DRIC Defence Research Information Centre (UK).

DRIDAC drum input to digital automatic computer.

DRIVE document read information verify and edit.

DRL 1. data requirements language. 2. direct retrieval language.

DRMS data resource management system.

DRO 1. destructive read out. 2. digital read out. Computing.

DROM decoder read-only memory.

DROP *Distribution register of organic pollutants*. Databank, *EPA*.

DROS disc resident operating system.

DRPS 1. disc real-time and programming system. 2. dynamic memory relocation and protection system, *SEMIS*.

DRQ data request.

DRS 1. data retrieval system. 2. direct relay satellite. Communications satellite. 3. distributed resource system, *ICL*. 4. document retrieval system.

DRT device reference table.

DRTM disc real-time monitor.

DRU data reference unit.

DRUGDOC selective dissemination of information service, Excerpta Medica (Netherlands).

DS 1. data scanning. 2. data set. 3. debugging system. 4. decimal subtract. 5. define storage. 6. digital signal. 7. disc storage. 8. disc system. 9. distributed system. 10. division/station code. Searchable field, Dialog. 11. drum storage.

DSA 1. data set adaptor, computing. 2. dataroute serving area, TransCanada Telephone System/Computer Communications Group. 3. dial service assistance, telecommunications. 4. direct storage access. 5. distributed systems architecture, Honeywell.

DSB 1. data set block. 2. double sideband, telecommunications.

DSBEC double sideband emitted carrier, telecommunications.

DSB-SC double side band-suppressed carrier, telecommunications.

DSC 1. data set controller. 2. direct satellite communications. 3. district switching centre, telecommunications network.

DSCB data set control block. Standard control block.

DSCS Defense Satellite Communication System (US).

DSD 1. digital system design. 2. digital system diagram.

DSDD double sided, double density. Magnetic disc format.

DSDL data storage description language.

DSDS dataphone switched digital service, *AT&T*.

DSE 1. data set extension. 2. data storage equipment. 3. data switching exchange. 4. data systems engineering. 5. direct switching equipment. 6. direct switching exchange, telecommunications. 7. distributed systems environment, Honeywell.

DSF disc storage facility.

DSI digital speech interpolation. Telecommunications.

DSID data set identification.

DSL 1. data set label. 2. data structures language. 3. digital simulation language.

DSLO distributed systems licensing option, *IBM*.

DSM 1. disc space management. 2. Dutch State Mines. Originator of *TISDATA* database on physical properties of compounds.

DSMS document service management system, *CAS DDS*.

DSN distributed system network.

DSO data set optimizer.

DSOS data switch operating system.

DSP 1. distributed system program. 2. dynamic support program.

DSPLC display controller, microprocessor.

DSR 1. data set ready. Model signal. 2. device state register.

DSS 1. data switching system. 2. decision support system. Computer-based information system. 3. digital subsystem. 4. digital switching subsystem, telecommunications. 5. disc support system. 6. document storage system. Word processing system, *ICL*. 7. dynamic support system.

DSSD double-sided single density. Magnetic disc format.

DST disc storage terminal.

DSU 1. data service unit. 2. disc storage unit.

DSV digital sum variation, telecommunications. Sum refers to maxima and minima of coded signal.

DSW 1. data status word. 2. device status word.

DSX 1. digital signal cross-connection equipment, telecommunications. 2. distributed system executive, *IBM*.

DT 1. data tags. Searchable field, *NLM*. 2. data terminal. 3. data translator. 4. data transmission. 5. date of publication or contract. Searchable fields, Dialog *IRS*. 6. dial tone. 7. disc-tape. 8. document title. Searchable field, *ESA-IRS*. 9. document type. Searchable field, Dialog and *SDC*. 10. down-time.

DTB 1. Danmarks Tekniske Bibliotek. Originates *ALIS* and operates *CT*, *COMPENDEX* and *FSTA*, also information broker (Denmark). 2. decimal to binary. 3. dynamic translation buffer.

DTC Data Terminals and Communications Inc. (US).

DTCU data transmission control unit, Burroughs.

DTD data transfer done.

DTE data terminal equipment, *CCITT*.

DTF define the file.

DTG display transmission generator.

DTI 1. Department of Trade and Industry (UK). 2. display terminal interchange.

DTL diode transistor logic.

DTMF dual tone multifrequency signalling.

DTMS database and transaction management system, *IBM*.

DTN data transporting network.

DTP 1. daily transaction reporting. 2. data transfer protocol.

DTR 1. data terminal ready. 2. data transfer register. 3. disposable tape reel. 4. distribution tape reel. Magnetic tape package.

DTS data transmission system.

DTT 1. data transition tracking. 2. data transmission terminal.

DTTU data transmission terminal unit, Burroughs.

DTU 1. data transfer unit. 2. data transmission unit. 3. digital tape unit. 4. digital transmission unit.

DTV digital to television.

DUAL dynamic universal assembly language.

DUALABS Data Use Access Laboratories Inc. Produces *NCS*.

DUC *Distributable union catalog.* Microfiche, Harvard University (US).

DUM disc user multi-access unit.

DUNMIRE Dundee University numerical method information retrieval experiment (UK).

DUO *DOS* under *OS*: disc operating system under (control of system's) operating system.

DUP disc utility program, *IBM*.

DUS diagnostic utility system.

DUT device under test.

DUV data under voice. Data transmission on voice grade channel.

DVA DiscoVision Associates. Videodisc manufacturer (US).

DVM 1. digital volt meter. 2. displaced virtual machine.

DVR design and verification routine, Sperry Univac.

DVS Deutscher Verband für Schweisstechnik. Co-originator of *Dokumentation schweisstechnik* database (FRG).

DVST direct view storage tube, Princeton Electronic Products (US).

DW 1. daisy wheel, printer. 2. double word, computing.

DWM destination warning marker, indicating tape end.

DWSS data highway service system, Toshiba.

DX 1. distance. 2. duplex.

DXC data exchange control, Sperry Univac.

DX/RSTS document transmission/resource timesharing, *DEC*.

DXS data exchange system, *TI*.

DXS/OS *DXS*/operating system, *TI*.

DXS/ST *DXS*/statement translator, *TI*.

DXS/TL *DXS*/transaction language, *TI*.

DYANA dynamics analyzer programmer. Computer program.

DYNASAR dynamic systems analyzer, General Electric (US).

DYSAC digitally simulated analog computer.

DYSTAC dynamic storage analog computer.

DYSTAL dynamic storage allocation language.

DZF Dokumentationszentrale Feinworktechnik. Originator, operator and its database on mechanical and light engineering (FRG).

E

E erlang. Unit of telecommunications traffic intensity.

EAB European American Bank. Originates *EURABANK* and *FECS*.

EABS *Euro-abstracts*. Database, *CEC*.

EAE extended arithmetic element.

EAI *Economic abstracts international*. Database, Akzo Zout Chemie Nederland BV and *LI*.

EAL Electronic Associates Limited.

EAM 1. electrical accounting machine. 2. evanescent access method, Sperry Univac.

EAN European article numbering. Machine readable labelling.

EAOS easy access ordering system. Automated book ordering system, Blackwells North America.

EAPROM electrically alterable programmable read-only memory.

EAROM electrically alterable read-only memory.

EARS *Epilepsy abstracts* retrieval service. Database and originator (US).

EAS extended area service, telecommunications.

EASE engineering automatic system for solving equations, General Motors.

EASY efficient assembly system. Assembler language.

EAX electronic automatic exchange. Stored program control switching, *GT & E*.

EBA European Business Associates. Information systems marketing organization.

EBAM electron beamed access memory.

EBCDIC extended binary coded decimal interchange code.

EBDIK *EBDIC* for Kana characters.

EBIB *Energy bibliography and index*. Database, Centre for Energy and Mineral Resources, Texas A & M University (US).

EBIS *Economic Information Systems business information system*. Database, *EIS*.

EBM extended branch mode.

EBR electron beam recording, on storage device.

EBU European Broadcasting Union.

EC 1. element count. Searchable field, Dialog *IRS*. 2. engineering change. 3. error correcting. 4. event code. Searchable field, Dialog *IRS*. 5. event count. Searchable field, *SDC*. 6. event counter.

ECAB *Economic abstracts*. Database, *BELINDIS*.

ECAP electric circuit analysis program.

ECAT European Centre for Automatic Translation (Luxembourg).

ECB event control block, computing.

ECC error checking and correction.

ECDIN *Environmental chemicals data and information network*. Databank, Euratom (Italy).

ECER *Exceptional child information resources*. Database, *CEC* (US).

ECHO European Commission Host Organization.

ECL 1. emitter coupled logic. Integrated circuit system. 2. Eurotec Consultants Ltd. Computer consultants (UK). 3. executive control language.

ECLA Economic Commission for Latin America. Originator of *Clades* database (UN).

ECLAT European Computer Leasing and Trading Association.

67

ECM 1. effective calls meter, telecommunications. 2. extended core memory.

ECMA European Computer Manufacturers' Association.

ECO 1. electronic central office, within network. 2. electronic contact operate, computing.

ECODU European Control Data Users Group.

ECOM electronic computer originated mail. *USPO* service.

ECOMA European Computer Measurement Association.

ECPS extended control program support, *IBM*.

ECS 1. European communications satellite. 2. experimental communications satellite (Japan). 3. extended core storage.

ECSA European Computing Services Association.

ECT electronic and control technology, *MEP*.

ECTR extended connection table representation. Chemical structure representation technique used in *IR*.

E-cycle execution cycle. Computing.

ED 1. edition. 2. entry date. Searchable field, *BLAISE* and *NLM*. 3. *ERIC* document. 4. errata data. Searchable field, Dialog *IRS*. 5. error detecting. 6. expanded display. 7. external device.

EDA electronic differential analyser.

EDAC error detection and correction.

EDB *Energy database*. Originated by *ORNL*.

EDBD *Environmental database directory*. Database, *NODC*.

EDBS educational database system, Computer System Research Group, University of Toronto (Canada).

EDC 1. error detection and correction. 2. external disc channel. 3. external drum channel.

EDD electronic document delivery.

EDE *Environmental data and ecological parameters*. Database, International Society for Ecological Modelling.

EDF Électricité de France. Database originator.

EDF-DOC *EDF – documentations*. Database, *EDF*.

EDGE electronic data gathering equipment.

EDICT engineering document information collecting technique.

EDIS elektronisches Dokumentations und Informationssystem. Information retrieval system, *ETH* (FRG).

EDMA extended direct memory access.

EDMS extended data management system, Xerox.

EDOS-MSO extended disc operating system – multistage operations, Fujitsu (Japan).

EDP 1. educational data processing. 2. electronic data processing.

EDPE electronic data processing equipment.

EDPIE *Eurodial principe indicateur économique*. Databank originated by *OECD*, Telesystems-Eurodial.

EDPM electronic data processing machine.

EDPS electronic data processing system.

EDRS *ERIC* document reproduction service.

EDS 1. electronic data system, government of FRG. 2. electronic document storage systems. 3. *Energy data system*. Databank, *EPA*. 4. exchangeable disc stores.

EDSAC electronic delay storage automatic calculator. One of the earliest computers, 1949.

EDSTAT *Educational statistics information access service*. Databank, *NCES*.

EDUCOM Educational Communications. Inter-university council (US).

EDUG European Datamanager Users Group.

EDUNET education network. Referral and directory system of *EDUCOM*.

EDVAC electronic discrete variable computer. One of first computers, 1949.

EDX event driven executive, *IBM*.

EEA Electronic Engineering Association (UK).

EEC 1. English Electric Computers (UK). 2. European Economic Community. See also *CEC*.

EEDB *ERDA energy database*. Database, *ERDA*.

EEL exclusive exchange line, telecommunications.

EEROM electronically erasable read-only memory. Type of computer storage.

EET equipment engaged tone, telecommunications.

EF 1. electronic filing. 2. error free region. 3. extended facility, *IBM*.

EFA, Wharton Wharton Economic Forecasting Associates. Originator and databank (US).

EFI error free interval.

EFT electronic funds transfer.

EFTP error file teaching package.

EFTS electronic funds transfer system.

EG expenditure greater than. Searchable field, Dialog *IRS*.

EGIF equipment group interface.

EGO *Économie/gouvernements/organisations*. Database, Ministère des Affaires Étrangères (France).

EHF extremely high frequency. Approximately 10^{11} *Hz*.

EHIS *Emission history information system*. Databank, *EPA*.

EHOG European Host Operators Group. Federation of 28 hosts.

EHV extra high voltage.

EIA Electronic Industries Association (US).

EIAJ Electronics Industry Association of Japan.

EIASM European Institute for Advanced Studies in Management. Originates and operates *Database on European doctoral theses in management*.

EIC Environment Information Center. Database originator (US).

EIES electronic information exchange system. Computer network (US).

EIII Association of the European Independent Information Industry.

EIMET engineering information meetings.

EIN European informatics network. Coordinated by *NPL*.

EIRB European Investment Research Bureau. Produces *Europrospects* database.

EIRP effective isotropically radiated power. Measure of power supply to antenna and gain, also *ERP* (US).

EIS 1. Economic Information Systems Inc. Database originator (US). 2. extended instruction set, Honeywell.

EIS Plants *Economic Information Systems – plants*. Databank on industrial plant, *EIS*.

EISs *Environmental impact statements*. Database, Heiner and Co. (US).

EIU Economist Intelligence Unit (UK).

EJCT *Engineering Joint Council thesaurus* (US).

EL expenditure less than. Searchable field, Dialog *IRS*.

ELAN elementary language. Programming language.

ELAS electronisches ausleihkontrol system. Library circulation control system, *ETH*.

ELCOM *Electronics and computers*. Database, Cambridge Scientific Abstracts (US).

ELD edge lit display.

ELE equivalent logic element.

ELECOMPS *Electronic components*. Databank, *ESA*.

ELF extremely low frequency. Radio communications.

ELHILL Lister Hill (founder of *NLM*). Information retrieval software, *NLM*.

ELI English language interpreter. Project on texts.

ELIAS *Environmental libraries automated system.* Database on environmental science and issues, Environment Canada.

ELMIG Electronic library Membership Initiative Group, *ALA*.

ELMS experimental library management system, *IBM*.

ELP element processor.

ELSIE 1. electronic letter sorting and indicating equipment. 2. electronic speech information system. Speech synthesis system, eg for electronic bus stop.

ELSPECS *Electronic specifications.* Databank of specifications issued by national agencies, *ESA*.

EM 1. electronic mail. 2. end of medium. 3. *Excerpta medica.* Database on medical sciences, Elsevier (Netherlands).

EMA *Equipment market abstracts.* Database, Predicast Inc. (US).

EMBASE *Excerpta medica database*, Elsevier.

EMC extended multiplexer channel.

EMI electromagnetic interference.

EMIC *Environmental mutagen information center.* Databank, *NIEHS*.

EMIS 1. electronic markets and information systems. Computerized chemical trading system, McGraw-Hill and I. P. Sharp. 2. electronic materials information service. Information retrieval service, *INSPEC*.

EML Economic Models Ltd. Originates *DIADEM* (UK).

EMMA extra *MARC* material. Material for which no *MARC* record exists.

EMMS electronic mail and message systems.

EMOL *Excerpta medica* online. Online *EM* search service.

EMS 1. Einheitliches Microfilm System. Standardised microfilm system (GDR). 2. electromagnetic spectrum. 3. electronic mail system. 4. electronic message system.

EMSS emergency manual switching system, telecommunications.

EMU emulator. Microprocessing.

EN 1. equipment number. 2. event name. Searchable field, Dialog *IRS*.

ENDS European nuclear documentation system, *EURATOM*.

ENEIDE *Ensemble normalisé sur les entreprises industrielles pour le développement économique.* Databank, *STISI*.

ENG electronic news gathering. Use of portable video cameras with transmission to studio for immediate broadcasting.

ENIAC electronic numerical integrator and calculator. One of the earliest computers, 1949.

ENQ enquiry. Transmission control character.

EN RX enable receive. Status activation code.

ENSDF *Evaluated nuclear structures data file.* Databank, *INKA*.

EN TX enable transmit. Status activation code.

EOA end of address.

EOE error and omission excepted.

EOF end of file.

EOI 1. end of information. 2. end of inquiry.

EOJ end of job.

EOL end of life.

EOM end of message.

EOR 1. end of record. 2. end of reel. 3. end of run.

EOT 1. end of tape. 2. end of task. 3. end of transmission.

EOV end of volume.

EP emulation program, *IBM*.

EPA 1. *Electronic publishing abstracts.* Database, *PIRA* and Pergamon. 2. Environmental Protection Agency (US). Databank originator.

EPAM elementary perceiver and memorizer. University of California system (US).

EPASYS European patents administration system. Information system, *EPO*.

EPB *Environmental periodicals bibliography.* Database, Environmental Studies Institute, International Academy at Santa Barbara (US).

EPC editorial processing centre.

EPCOT experimental prototype community of tomorrow, Walt Disney, Florida (US).

EDPT estimated project duration time.

EPIA *Electric power industry abstracts.* Database, Edison Electrical Institute (US).

EPIC 1. electronic printer image construction. Xerox software for graphics output. 2. *Electronic properties information center.* Databank, *CINDAS*. 3. *Estimation de propriétés pour l'ingénieur chimiste.* Databank, Université de l'Etat à Liège (Belgium). 4. *Exchange price indicators.* Database, London Stock Exchange (UK).

EPL encoder programming language.

EPO European Patent Office.

EPOS electronic point of sale.

EPR error pattern register.

EPRI Electronic Power Research Institute (US). Originates *EPRI RDS*.

EPRI RDS *EPRI research and development information system.* Databank, *EPRI*.

EPROM 1. electrically programmable read-only memory. 2. erasable programmable read-only memory.

EPS even parity select.

EPSS experimental packet switching service (UK).

EPU executive processing unit.

EQ enquiry.

ER error recovery.

ERA Electrical Research Association (UK).

ERCC 1. Edinburgh Regional Computing Centre. 2. error checking and correction.

ERD *Energy research and development inventory.* Databank, *ORNL*.

ERDA Energy Research and Development Agency. Database originator (US).

ERES environmental record editing and statistics, Fujitsu (Japan).

ERFPI extended range floating point interpretive system.

ERGODATA *Ergonomy data.* Database, Laboratoire d'Anthropologie et d'Ecologie Humaine, Université Réné Descartes, Paris (France).

ERIC *Educational resources information center.* Set of databases of US Office of Education.

ERISTAR earth resources information storage transformation analysis and retrieval. Auburn University for *NASA* (US).

ERJE extended remote job entry.

ERL echo return loss, *VNL*.

EROICA *Estimation and retrieval of organic properties.* Databank, University of Tokyo (Japan).

ERP 1. effective radiated power. See *EIRP*. 2. error recovery procedure.

ERT expected run-time.

ES 1. edinaya sistema (uniform system). Computer series (USSR). 2. element signal. Searchable field, Dialog *IRS*. 3. establishment data. Searchable field, Dialog. 4. external store.

ESA 1. Entomological Society of America. Co-originator of *Pesticides index* databank. 2. European Space Agency. Runs host *IRS* and communications satellite programme. 3. externally specified address.

ESA-IRS European Space Agency information retrieval service. Host for mainly scientific databases.

ESANET *ESA* network. Provides access to *ESA-IRS*.

ESC 1. Economic Sciences Corporation. Originator and its databank (US). 2. escape. Control character.

ESCAP Economic and Social Commission for Asia and the Pacific. *UN* division originating *PCHIS*.

ESCAPE expansion symbolic compiling assembly program for engineering.

ESCS emergency satellite communications system.

ESD 1. electrostatic storage deflection. 2. end sequence done.

ESFI epitaxial silicon film on insulator, *NMOS*.

ESI 1. École des Sciences de l'Information. *UNESCO* initiative (Morocco). 2. *Environmental science index.* Database, Environmental Information Center Inc. (US). 3. Essex International. Microprocessor manufacturer (US). 4. externally specified index.

ESIC Environmental Science Information Center, US National Oceanic and Atmosphere Administration.

ESIS *European shielding information service.* Databank, *EURATOM*.

ESIT Egyptian Society for Information Technology.

ESP externally supported processor. Mainframe computer from Software AG.

ESPES *Especialidades farmacéuticas Espanoles.* Databank on drug information, *CINIME*.

ESPL electronic switching programming language.

ESPOL executive system problem oriented language, Burroughs.

ESPRIT European strategic programme on research in information technology.

ESR electronic send receive.

ESRO European Space Research Organisation, *ESA*.

ESS 1. echo suppression system, telecommunications. 2. electronic switching system, telecommunications. 3. event scheduling system.

ESTC *Eighteenth-century short title catalogue.* Database, *BL*.

ESTV error statistics by tape volume.

ESU electrostatic unit.

ET 1. end of text. 2. engage tone, telecommunications. 3. engage test, manual exchanges, telecommunications. 4. English title. Searchable field, *ESA-IRS*.

ETB end of transmission block.

ETC 1. Electronic Tool Company (US). 2. extended text compositor, *ADR* (US). 3. European Translations Center (Netherlands).

ETGTS electronic text and graphics transfer system.

ETH Eidgenossische Technische Hochschule. Originates *Chemco* databank (Switzerland).

ETIC *Environmental technology information center.* Database, *NIEHS*.

ETIM elapsed time.

ETIS European Technical Information Service. Information broker and database originator.

ETIS-MARFO *ETIS in machine readable form.* Database, *ETIS*.

ETL educational technology language, University of Western Ontario (Canada).

ETMF elapsed time multiprogramming factor.

ETP electronic tough pitch. Grade of copper used as conductor.

ETR expected time of response.

ETS 1. electronic tandem switching. 2. engineering test satellite (Japan). 3. enquiry terminal system.

ETSS entry time sharing system, *IBM*.

ETV educational television.

ETX end of text.

EU execution unit.

EUF end user facility.

EUPEPTIC evaluation of unitary programs for effecting plural tasks in index construction.

EURABANK *European-American bank.* Databank on activities of non-US banks, *EAB*.

EURATOM European Atomic Energy Community. Originates and operates *EUROCOPI*.

EURIM European Conference on Research into Management of Information.

EURIPA European Information Providers Association.

EURIS European Information Service. Host for *CELEX* (Belgium).

EUROBASE *European database*. Databank on election results, *INFAS*.

EUROCOPI European computer program information centre. Databank, *EURATOM/JRC*.

EURODICAUTOM European automated dictionary. Machine translation system, *CEC*.

EURONET European network, for online searching, administered by *CEC*.

EUSIDIC European Association of Scientific Information Dissemination Centres. Promotion organization.

EUSIREF 1. European Association of Science Information Referral Centres. 2. European Scientific Information Retrieval Working Group, *EUSIDIC*.

EUTELSAT European telecommunications satellite organization. Established by European *PTT*s to operate space segment of *ECS*.

EUV extreme ultra-violet, telecommunications.

EV errata volume. Searchable field, Dialog *IRS*.

EVA error volume analysis.

EVDS electronic visual display subsystem.

EVE data entry and validation equipment, Logical Machine Corporation (US).

EVIL extensible video interactive language.

EVK evaluation kit, American Microsystems Inc.

EVM extended virtual machine.

EVPI expected value of perfect information.

EVR electronic video recording.

EWF *Elektronisches Wörterbuch der Fachsprachen*. Multilingual terminology bank, Technische Universität Dresden (GDR).

EX 1. exchange. 2. expenditure. Searchable field, Dialog *IRS*.

EXCP execute channel program.

EXDAMS extendable debugging and monitoring system.

EXDC external data controller.

EXEC 1. execute. 2. executive system. 3. operating system, not an acronym.

EXNOR exclusive *NOR*. Logical operation or electronic gate.

EXOR exclusive *OR* gate. Logical operation or electronic gate.

EXPIO expander input/output, microprocessing.

EXR execute and repeat.

EY entry year. Searchable field, *SDC*.

F

F 1. farad. Unit of capacitance. 2. femto. 10^{-15}. 3. final, as subscript. 4. fixed head.

FA 1. field address. 2. *Forskning og utvikling indeks fast avfall*. Environmental database, *NSI*.

FAC 1. file access channel. 2. floating accumulator.

FACE 1. field alterable control element 2. field artillery computer equipment.

FACES *FORTRAN* automated code evaluation system.

FACS floating decimal abstract coding system.

FACT 1. Fast access current text bank. Electronic library, University of Missouri (US). 2. Federation against Copyright Theft. International organization on videotape piracy. 3. Foundation for Advanced Computer Technology (US). 4. fully automated cataloging technique. Document control system, Library Micrographic Services Inc. (US). 5. fully automated compiling technique. Honeywell system. 6. fully automatic compiler translator.

FACTS facsimile transmission system. Experimental project (US).

FAD floating *AND*, computing.

FADAC field artillery digital automatic computer.

FADS *FORTRAN* automatic debugging system.

FAHQMT fully automatic high quality machine translation.

FAHQT fully automatic high quality translation.

FAIR *Fast access information retrieval*. Database on biomedical engineering, *MRC*.

FAIRS 1. Federal Aviation information retrieval system (US). 2. Food and Agriculture Organization agricultural information storage and retrieval system. Operated by *FAO*. 3. fully automated information retrieval system.

FAKS file access keys.

FAL file access listener.

FALTRAN *FORTRAN* to *ALGOL* translator.

FAM 1. fast access memory. 2. file access manager, Prime Computers.

FAMIS factory management information system, *BL Systems* (UK).

FAMOS 1. fast multi-tasking operating system. 2. floating gate avalanche injection metal oxide semiconductor. Storage element, *EPROM*.

FAMS forecasting and modelling system.

F and S 1. *Fast and systematic*. Set of databases, Predicasts Inc. (US). 2. Frost and Sullivan Inc. Originate *DMS* (US).

FAO Food and Agriculture Organization. *UN* body, database and databank originator and operator.

FAP 1. failure analysis program. 2. floating point arithmetic package. 3. floating point arithmetic system, Lockheed. 4. *FORTRAN* assembly program.

FAPNEWDT financial accounts package new data. Financial accounting software, Torch Computers Ltd.

FAQS fast queuing system.

FAR 1. failure analysis report. 2. file address register.

FARMDOC pharmaceutical documentation. Patents retrieval system, Derwent Publications (UK).

FAS Fuel availability system. Databank on coal, *BOM*.

FASE fundamentally analysable simplified English.

FASIT fully automatic syntactically-based indexing system.

FAST 1. fast access storage technology. 2. flexible algebraic scientific translator, *NCR*. 3. formula and statement translator.

75

FAT formula assembler translator.

FATAR fast analysis tape and recovery.

FAUL Five Associated University Libraries. Cooperative, New York State (US).

FAUST folkebibliotekernes automation system. Public libraries automation system (Denmark).

FAX facsimile transmission.

FAXCOM facsimile communication service, TransCanada Telephone System.

FB 1. file block. 2. fixed block.

FBA fixed block architecture.

FBM foreground and background monitor.

FBR 1. Forskningsbiblioteksradet (Swedish council for research libraries). Co-originates *LIBRIS*. 2. full bibliographic record.

FC 1. ferrite core. 2. file code. 3. file conversion. 4. font change. Typesetting. 5. The Foundation Center. Produces databanks on grant awarding foundations (US). 6. foundation city. Searchable field, Dialog *IRS*. 7. free cursor. 8. function code.

FCB 1. file control block. 2. forms control buffer. 3. function control block.

FCC 1. Federal Communications Commission. Telecommunications regulatory body (US). 2. frame check character.

FCCTS federal *COBOL* compiler testing service.

FCD *Fine chemicals directory*. Databank, Fraser Williams Scientific Systems (UK).

FCFO full cycling file organization.

FCFS first come first served.

FCI flux change per inch.

FCL 1. feedback control loop. 2. format control language.

FCM firmware control memory.

FCP 1. file control package. 2. file control processor. 3. file control program. 4. Foxboro Control Package, Foxboro (US).

FCR *Federal court reports*. Database, Department of Justice, Canada.

FCS 1. facsimile communications system. 2. file control services. 3. file control system. 4. fixed control storage. 5. frame check sequence.

FCU file control unit.

FCUS *FORTRAN* compiler validation system.

FD 1. file definition. 2. file description. 3. file directory. 4. flexible disc. 5. floppy disc. 6. frequency demodulator. 7. full duplex.

FDA Food and Drug Administration. Database and databank producer (US).

FDB 1. file data block. 2. field descriptor block.

FDC 1. facsimile data converter. Facilitates communication between facsimile terminal and computer. 2. floppy disc controller.

FDCS functionally distributed computing system.

FDD 1. flexible disc drive. 2. floppy disc drive.

FDDL field data description language.

FDEP formatted data entry program, Mohawk Data Systems.

FDIC Federal Deposit Insurance Corporation. Originator and its databank (US).

FDM frequency division multiplexing. Telecommunications.

FDMA frequency division multiple access. Telecommunications.

FDMS floppy disc management system, LogAbax.

FDOS floppy disc operating system, LogAbax.

FDP 1. fast digital processor. 2. Foxboro display packages.

FDR file data register.

FDS 1. fixed disc stores. 2. flexible disc system. 3. floppy disc system. 4. frequency division separator, multiplexing.

FDSR floppy disc send receive.

FDT 1. formatted data tapes. 2. functional description table.

FDU flexible disc unit.

FDV fault defect verification.

FDX full duplex.

FE 1. format effector. Control character. 2. front end.

FEA Federal Energy Administration. Originates *Energy* databank (US).

FEAT frequency of every allowable term.

FEB functional electronic block, integrated circuits.

FEC 1. floating error code. 2. forward error correction, telecommunications.

FECP front end communications processor.

FECS *Foreign exchange counselling system*. Databank, *EAB*.

FEDD for early domestic dissemination. *NASA*.

FEDEX *Federal index*. Database covering federal government activities, Capitol Services International (US).

FEDLINK Federal library and information network. Computer network (US).

FEDNET Federal network. Computer network (US).

FEDREG *Federal register*. Database on federal government proclamations etc., Capitol Services International (US).

FEDS fixed and exchangeable disc storage.

FED-STD federal standard (US).

FEFO first ended first out. Processing procedure.

FEM firmware expansion model, *HP*.

FEP 1. financial evaluation program. 2. *FORTRAN* enhancement package, Interdata. 3. front end processor.

FES forms entry system, Basic Four (US).

FET field effect transistor.

FEXT far-end crosstalk, telecommunications.

FF 1. flip flop. Electronic circuit. 2. form feed. Control character.

FFM fast file manager.

FFT fast Fourier transformation. Noise reduction technique in transmitted signal.

FG funding greater than. Searchable field, Dialog.

FGC fifth generation computer.

FHD fixed head disc.

FHLBB Federal Home Loan Bank Board. Originates *FSLIC* (US).

FHSF fixed head storage facility.

FIB 1. *Fachinformationsbank*. Databank (FRG). 2. file information block.

FID Fédération Internationale de Documentation.

FIDAC film input to digital automatic computer. Georgetown University (US).

FIDACSYS *FIDAC* system.

FIDAS formularorientiertes Interaktives Datenbanksystem. *IRS* (FRG).

FID/OM *FID* committee on operational machine techniques and systems.

FID/TM *FID* committee on theory and method of systems, cybernetics and information networks.

FID/TMO *FID* committee on theory, methods and operation of information systems and networks.

FIFO 1. first in first out. Queue discipline, computing and telecommunications. 2. floating input/floating output.

FILEX file exchange.

FILO first in last out. Processing procedure.

FILU four bit interface logic unit.

FINAC fast interline non-active automatic control.

FIND 1. File of Industrial Data. Originator and database (UK). 2. file interrogation of nineteen hundred data, *ICL*.

FINFO first in not used first out. Processing procedure.

FINTEL *Financial Times electronic publishing*. Group of databases on business information, Financial Times (UK).

FIOP *FORTRAN* input/output package.

FIPS federal information processing standard.

FIPSCAC *FIPS* Coordinating and Advisory Committee.

FIR file indirect register.

FIRL faceted information retrieval system for linguistics.

FIRST 1. fast interactive retrieval system technology. 2. federal information research science and technology network, *COSATI*.

FIS Fachinformationssystem. Information system (FRG).

FISHROD fiche information selectively held and retrieved on demand. Wellcome Foundation (UK).

FIT file inquiry technique.

FIU Federation of Information Users (US).

FIX fault isolater and exercizer, Honeywell.

FIXIT flexible information exploitation interpretive transfer. Software engineering tool.

FIZ Fachinformationszentrum. Information centre (FRG).

FIZ-technik Fachinformationszentrum Technik. Host (FRG).

FJCC Fall Joint Computer Conference (US).

FL 1. fault location. 2. field length. 3. focal length. 4. funding less than. Searchable field, Dialog *IRS*.

FLA Francois Libman Associates. Online information broker (France).

FLAG *FORTRAN* load and go, Xerox.

FLAIR *FORTRAN* language in core rapid translator, Xerox.

FLDEC floating point decimal.

FLF fixed length field.

FLI free language indexing. Information retrieval.

FLIM fast library maintenance.

FLINT floating interpretive language.

FLIP 1. floating indexed point arithmetic. 2. floating point interpretive program.

FLIR forward looking infrared.

FLIRT 1. Federal Librarians Round Table, *ALA*. 2. free language information retrieval tool. Institut voor Wiskunde Informatieverwerking en Statistik (Netherlands).

FLIT 1. fault location by interpretive testing. 2. flexowriter interrogation tape.

FLITE *Federal legal information through electronics*. Legal database (US).

FLMEM floppy disc memory.

FLNPP Federal library network prototype project (US).

FL/1 function language one.

FLOP floating octal point.

FLOTRAN *flowcharting FORTRAN*.

FLPAU floating point arithmetic unit.

FLPDC floppy disc controller.

FLPL *FORTRAN* list processing language.

FLS free line signal, telecommunications.

FLT fault location technology.

FLTSATCOM fleet satellite communication system (US).

FM 1. file maintenance. 2. file management. 3. frequency modulation.

FMDU fast multiply/divide unit. Form of *ALU*.

FMIS fiscal management information system.

FML file manipulation language.

FMLF file management loading facility.

FMPP flexible multipipeline processor.

FMPS functional mathematical programming system

FMS 1. file management supervisor. 2. file management system. 3. financial management system. Financial Management and Software Ltd. 4. flexible manufacturing system. Use of robots in manufacturing.

FN 1. foreign patent number. 2. foundation name. Searchable fields, Dialog *IRS*. 3. functional network.

FNA Fujitsu network architecture (Japan).

FNB Fédération Nationale du Bâtiment. Originates and operates *ARIANE* database and databanks (France).

FNP front end network processor. Auxiliary computer for network operations.

FNR file next register.

FNTL *FINTEL*.

FOA Forsvarets Forskningsanstalt (Swedish National Defence Research Institute). *NTIS* operator.

FOC fibre optics communications.

FOCAL formula calculator.

FOD functional operational design.

FOIA Freedom of Information Act (US).

FOIL file oriented interpretive language.

FOM fibre optic modem.

FORAST formula assembler and translator. Programming language.

FORC formula coder. Programming language.

FORCE *FORTRAN* conversational environment.

FORDAP *FORTRAN* debugging aid program.

FOREM file organization evaluation model.

Forest AIDS *(Forest products) abstract information digest service*. Database, *FPRS*.

FORGE file organization generator.

FORIMS *FORTRAN* oriental information management program.

FORMAC formula manipulation compiler.

FORS *Forschungsprojekte aus raumordnung, stadtebau und wohnungswesen*. Databank on housing and urban planning, *IRB*.

FORTRAN formula translation. Programming language.

FOSDIC film optical scanning device for input to computer.

FOT fréquence optimum de travail/optimum traffic frequency, telecommunications.

FOTS fibre optics transmission system.

FOU *Forskning og utvikling indeks*. Multi-disciplinary database, *NSI*.

4 PSK *QPSK*.

FP 1. file protect. 2. fixed point. 3. floating point. 4. function processor.

FPA floating point arithmetic.

FPGA field programmable gate array.

FPH floating point hardware.

FPL 1. field processing language, Olivetti. 2. Fox programming language.

FPLA field programmable logic array.

FPM 1. file protect memory. 2. floppy disc processor module, Transdata. 3. frames per minute. *TV* transmission standard.

FPP 1. fixed point protocol. 2. floating point process. 3. floating point processor.

FPQA fixed portion queue area.

FPR floating point register.

FPROM field programmable *ROM*.

FPRS Forest Products Research Society. Originator of *Forest AIDS* database (US).

FPS frames per second.

FPSK frequency and phase shift keying.

FPU floating point unit.

FQL formal query language.

FQS friendly query system, *IBM*.

FR 1. federal region. Searchable field, Dialog *IRS*. 2. file register. 3. forced release, of circuit. 4. foreign. Searchable field, *SDC*.

FRA Federal Radio Act (US).

FRANCIS 1. *Fichier de recherches automatisées sur les nouvautés, la* *communication et l'information en sciences sociales et humaines.* Social science database, *CNRS.* 2. *Food Research Association Computerized Information Service.* Database, Food Research Association (US).

FRB faultsman's ring back, telecommunications.

FRC Federal Radio Commission (US).

FR-DLP frame recognition – data link processor.

FRED 1. fractionally rapid electronic device. 2. front end for databases. Online searching system, General Telephone and Electronics Laboratories (US).

FRESS file retrieval and editing system. Early video editing system.

FRI flux reversals/inch. Magnetic storage measure.

FRIL fuzzy relational inference language. Systems language.

FRIMP flexible reconfigurable interconnected multiprocessor.

FRINGE file and report information processing generator. *GEC* system.

FRMM flux reversals/millimetre, of magnetic storage medium.

FROM fusible *ROM*.

FRPS flux reversals per second, of magnetic storage medium.

FRR functional recovery routines.

FRXD fully automatic reperforator-transmitter, telecommunications.

FS 1. field of science. Searchable field, Dialog *IRS.* 2. file segment. Searchable field, *SDC.* 3. file separator. Control character. 4. final selector, telecommunications. 5. foundation state. Searchable field, Dialog *IRS.* 6. function select. 7. fusible link. *PROM* programming technique.

FSCB file system control block.

FSCR field select command register.

FSD full scale deflection.

FSEC Federal Software Exchange Center (US).

FSHDB *World fishing catch database.* Databank, *FAO.*

FSK frequency shift keying. Data transmission.

FSL formal semantic language, *MIT* (US).

FSLIC *Federal savings and loan insurance.* Databank, *FHLBB.*

FSOS free-standing operating system, General Automation.

FSP frequency shift pulsing.

FSR file storage region.

FSS flying spot scanner, in *CRT*

FST file status table.

FSTA *Food science and technology abstracts.* Database, *IFIS.*

FSU 1. facsimile switching unit. 2. field select unit. 3. final signal unit, in *CCS.*

FT 1. foundation type. Searchable field, Dialog *IRS.* 2. French title. Searchable field, *ESA-IRS.* 3. functional test.

FTA fault tree analysis.

FTC fault tolerant compiler.

FTET full-time equivalent terminals, computer performance.

FTF fault transfer facility.

FTI *Foreign traders index.* Databank, *DOC.*

FTL fast transient loader.

FTP file transfer protocol.

FTPI flux transitions per inch.

FTR full text retrieval.

FTS 1. Federal Telecommunications System (US). 2. free time system. 3. Future Technology Systems, Scotland (US).

FTSC Federal Telecommunications Standards Committee (US).

FTU first time use.

FU funding. Searchable field, Dialog *IRS.*

FUINCA Fundación de la Red de Información Cientifica Automatizada. Database producer (Spain).

FUJ Fujitsu (Japan).

FUND funding. Searchable field, *SDC*.

FUNLIS fundamentals of library and information science, Drexel University (US).

FUS *FORTRAN* utility system.

FVU file verification utility.

FW first word.

FWA first word address.

FWL fixed word length.

FWRS *Fish and wildlife reference service.* Database, Denver Public Library (US).

FWS US Fish and Wildlife Service. Database originator.

FX 1. fixed area, of magnetic disc. 2. foreign exchange, telecommunications.

FX Database *Foreign exchange rates database.* Databank produced by Conticurrency (US).

FXT fixed time call, telecommunications.

FY fiscal year. Searchable field, Dialog *IRS*.

FYI *For your information.* Databank on finance, sport, weather etc., *WU*.

G

G giga. One thousand million (10^9).

GA 1. General Automation. Computer manufacturer (US). 2. global address.

Ga As gallium arsenide. Semiconductor material.

GAB graphic adaptor board.

GAELIC Grumman Aerospace engineering language for instructional checkout.

GALS military satellite, not an acronym (USSR).

GAM 1. *Globe and mail*. Newspaper databank (Canada). 2. graphic access method, *IBM*.

GAMA graphics assisted management application.

GAMIS general analytical methods information service, Laboratory of the Government Chemist (UK).

GAMS Groupement pour l'Avancement des Méthodes Spectroscopiques et Physio-chimiques d'Analyse. Originates and operates *CIS* databank (France).

GAN generating and assembly networks.

GAO General Accounting Office. Originates Congressional sourcebook series of databases (US).

GAP 1. general assembly program. 2. graphics application program.

GAPHYOR *Gaz-Physique-Orsay*. Database on properties of atoms, molecules, gases and plasmas, *CNRS* and Université de Paris XI (France).

GARF Graphic Arts Research Foundation (US).

GASP 1. general activity simulation program. Based on *FORTRAN*. 2. graphic applications subroutine package.

GASS generalized assembly system.

GAT 1. generalized algebraic translator. 2. graphic arts terminal, for phototype-setting.

GATD graphic analysis of three-dimensional data.

GATE generalized algebraic translator extended.

GATF Graphics Arts Technical Foundation (US).

GATS general acceptance test software.

GATT General Agreement on Tariffs and Trade. *UN* agency, originates and operates *TTDF*.

GATTIS Georgia Institute of Technology technical information service.

GB 1. gigabit. 10^9 bits. 2. gigabyte. 10^9 bytes.

GBF geographic base file.

GBLIC Gaussian band limited channel.

GBT generalized burst trapping.

GC 1. gigacycle. 2. group code. Searchable field, Dialog *IRS*.

GCAP generalized circuit analysis program.

GCCA Graphic Communications Computer Association (US).

GCE General Consumers Electronics (US).

GCH gigacharacters.

GCHQ Government Communications Headquarters (UK).

GCI 1. generalized communication interface. 2. graphics command interpreter. Provides plotting function on computer system.

GCOS 1. general comprehensive operating supervisor, Honeywell. 2. general comprehensive operating system.

GCPS gigacycles per second. 10^9 Hz.

GCR group code recording. Data storage method.

GCS 1. general communications system, Sperry Univac. 2. General Computer Systems (US). 3. graphic compatibility system, US Military Academy.

GCSG Graphic Communications Societies Group (UK).

GCT graphics communications terminal.

GD graphic display.

GDB global database.

GDBMS generalized database management system.

GDC General Data Comm (US).

GDDL graphical data definition language.

GDE 1. generalized data entry. 2. ground data equipment.

GDF group distribution frame, *FDM* system, telecommunications.

GDI graphic display terminal. Type of *VDU*.

GDL graphic display library.

GDM global data manager.

GDMS generalized data management system.

GDP goal directed programming.

GDS 1. graphic data system. 2. graphic design system.

GDSF generalized data structure definition facility.

GDU graphic display unit.

GE 1. gateway exchange, telecommunications. 2. General Electric. *IT* manufacturer (US). 3. geographic descriptor. Searchable field, Dialog *IRS*. 4. greater than or equal to. Relational operator.

GEC General Electric Company (UK).

GECOM general compiler.

GECOS general comprehensive operating supervisor. *GEC* multiprocessing system.

GE economic forecasts *General economic forecasts.* Databank, *GEISCO*.

GEEP General Electric electronic processor.

GEESE General Electric electronic system evaluator.

GEFRC general file/record control, Honeywell.

GEI General Instrument Corporation (US).

GEIS General Electric information services.

GEISA *Gestion d'études des informations spectroscopiques atmosphériques.* Databank on spectroscopy of planetary atmospheres, *CNRS*.

GEISCO General Electric Information Services Company. Division of General Electric and Honeywell conglomerate, databank originator and operator (US).

GEL general emulation language.

GELOAD general loader, Honeywell.

GEMCOS generalized message control system, Burroughs.

GEMT *Group of European metallurgical thermodynamicists.* Databank, National Physical Laboratory (UK), Thermodata (France), *RWTH* (FRG) and Vrije Universiteit (Belgium).

GENESYS generalized system. Computer program.

GENIE general information extractor.

GENSAL generic structure language, for chemicals information retrieval.

GEOARCHIVE *Geology archive.* Database on earth science, Geosystems (UK).

GEOREF *Geological reference file.* Database, American Geological Institute.

GEORGE general organizational environment. Operating system, *ICL*.

GEPAC General Electric process automation computer.

GERT graphical evaluation and review technique.

GERTS 1. general electric remote terminal supervisor, also *GRTS*, Honeywell. 2. general remote terminal system.

GESC Government Electronic Data Processing Standards Committee (Canada).

GESTA *Gesetzgebungsstand.* Database on progress in legislation, Bundestag (FRG).

GESYDAMS Groupe d'Exploitation des Systèmes de Documentation Automatisés en Médicine et en Science. Laval University (Canada).

GETEL General Electric test engineering language.

GEVIC General Electric variable increment computer.

GFI guided fault isolation.

GFM graphics function monitor, Tektronix.

GFP generalized file processor.

GFPBBD Groupement Francaise des Producteurs de Bases et Banques de Données. Database and databank producers association (France).

GFS *Government finance statistics.* Databank, *IMF.*

GG grant greater than. Searchable field, Dialog *IRS.*

GHz gigaherz. 10^9 Herz.

GI 1. *Galleries index.* File of *National foundation* database, *FC.* 2. General Instrument Corporation (US).

GIC 1. general input channel. 2. general input/output channel.

GID Gesellschaft für Information und Dokumentation. Host and centre for research in information science (FRG).

GIDEP Government-industry data exchange program (US).

GIEWS *Global information and early warning system.* Databank on food and agriculture, *FAO.*

GIFS generalized interrelated flow simulation.

GIFT general internal *FORTRAN* translator.

GIGO garbage in garbage out, computing.

GIM 1. General Instrument Microelectronics (UK). 2. generalized information management. 3. generalized information management language.

GIMIC guard ring isolated monolithic integrated circuit.

GINO graphical input/output.

GIOP general purpose input/output processor.

GIP general information programme, *UNESCO.*

GIPS ground information processing system.

GIPSY generalized information processing system, US Geological Survey.

GIR Stichting Gemeensschappelijke Informatieverwerking voor de Rundveehouder. Originates and operates agricultural databases (Netherlands).

GIRL 1. generalized information retrieval language. Search language, US Defense Nuclear Agency. 2. graph information retrieval language.

GIRLS generalized information retrieval and listing system.

GIS 1. generalized information system. 2. geographic information system. 3. Geo-science Information Society, American Geological Institute. 4. *Grant information system.* Databank, Oryx Press (US). 5. *Guidance information system.* Databank on education, Time Share Corporation (US).

GITIS Georgia Institute of Technology School of Information Science. Report series code (US).

GJP graphic job processor.

GKIII *General catalogue, 3rd edition, BLRD.*

GKS graphic kernel system. Graphics standard, *ISO.*

GL 1. geographic location. 2. grant less than. Searchable fields, Dialog *IRS.*

GMA *Geomechanics abstracts.* Database, Royal School of Mines (UK).

GMAP 1. general macro assembly program, Honeywell. 2. generalized macroprocessor.

GMIS generalized management information system.

GML 1. generalized markup language. 2. generic markup language. 3. graphic machine language.

GMP *Geometric modelling project. CAD* software, Glacier Metal, *ICI,* Northern Engineering Industries, J. H. Fenner and *DEC* (UK).

GMR general modular redundancy.

GMS general maintenance system.

GMSS graphical modelling and simulation system.

GMT 1. generalized multi-tasking. 2. graphics mouse technology. Uses hand-operated 'mouse' to instruct computer.

GN 1. grant number. 2. group number. Searchable fields, Dialog *IRS*.

GNC graphic numerical control.

GND ground. Earth connection.

GO generated output.

GOC graphic option controller, Sigma.

GOCI general operator computer interaction.

GOE ground operation equipment. Satellite operation system, Westinghouse.

GOL 1. general operating language. 2. goal oriented language.

GOLEM grosspeicherorientierte, listenorganisierte Ermittlungsmethode. *IR* software, Siemens.

GOM group occupancy meter, telecommunications.

GOMAC Government Microcircuit Applications Conference.

GOS 1. grade of service. 2. graphics operating system, Tektronix.

GP 1. gang punch. 2. general processor. 3. general purpose. 4. generalized programming. 5. Government Printing Office number. Searchable field, Dialog *IRS*. 6. graphics processor.

GPA 1. general purpose analysis. 2. general purpose array.

GPAX general purpose automation executive, *IBM*.

GPC 1. general peripheral controller. 2. general purpose computer.

GPCA general purpose communications adaptor.

GPD general protocol driver.

GPDC general purpose digital computer.

GPGL general purpose graphic language.

GPIA general purpose interface adaptor.

GPL 1. general purpose language. 2. general purpose loader.

GPLAN generalized database planning system.

GPLP general purpose linear programming.

GPM general purpose macro-generator.

GPO 1. General Post Office (UK). 2. Government Printing Office (US).

GPOS general purpose operating system, *TI*.

GPP general print and punch.

GPS general problem solver.

GPSDW general purpose scientific document writer.

GPSS 1. general problem statement simulator. 2. general purpose simulation system. High level language for *CAD*. 3. general purpose system simulator.

GPT general purpose terminal.

GQE generalized queue entry.

GR 1. general purpose register. 2. grade. 3. growth rate. Searchable fields, Dialog *IRS*.

GRA *Government report announcements*. Database, *NTIS*.

GRACE graphic arts composing equipment.

GRADS generalized remote access data base.

GRAFTEK Graphics Technology Corporation (US).

GRAIN graphics oriented relational algebraic interpreter.

GRAMLIN Greater Manchester local government information network (UK).

GRANADA grammatical non-algorithmic data description.

GRANIS graphical natural inference system.

GRAPDEN graphic data entry unit.

GRASP 1. generalized read and simulate program. 2. generalized remote acquisition and sensor processing. 3. graphic service program.

GR/CIDS *Genetic resources/communication, information and documentation system.* Databank, *FAO*.

GRED generalized random extract device.

GRG graphical rewriting grammar.

GRID graphical interactive display.

GRIN graphical input.

GRINDER graphical interactive network designer.

GRINS general retrieval inquiry negotiation structure. Lehigh University (US).

GRIP 1. general retrieval of information program. Personal indexing system, Hoechst Pharmaceutical Research Laboratories (UK). 2. Grandmet Information Processing (UK). 3. graphics interactive program.

GRIPHOS general retrieval and information processor for humanities oriented studies.

GRIPS general relation-based information processing system. Retrieval software, *DIMDI*.

GRISA Groupe de Recherche de l'Information Scientifique et Automatique, *EURATOM*.

GROATS graphical output package for *ATLAS*. Oriental and ancient language graphics plotting system, Atlas Computer Laboratories (UK).

GRP group reference pilot, telecommunications.

GRS 1. general register stack. 2. general reporting system. 3. generalized retrieval system.

GRTS 1. general electric remote terminal supervisor, also *GERTS*, Honeywell. 2. general remote terminal supervisor.

GS group separator. Control character, information separator.

GSA 1. general services administration. 2. general syntax analyser.

GSAM generalized sequential access method.

GSC group switching centre, telecommunications.

GSD General Systems Division, *IBM*.

GSDS general status display system. Graphics system, Rantek (US).

GSE graphics screen editor.

GSI 1. grand scale integration. Integrated circuit. 2. graphic structure input.

GSIS Group for the Standardization of Information Services (US).

GSL 1. generalized simulation language. 2. generation strategy language.

GSM graphics system module.

GSP 1. general syntactic processor. 2. graphics subroutine package.

GSR global shared resources.

GSS 1. Government Statistics Service (UK). 2. graphic support software.

GT 1. German title. Searchable field, *ESA-IRS*. 2. graphics terminal. 3. greater than. Relational operator.

G/T gain-to-noise temperature ratio.

GT & E General Telephone & Electronics. Common carrier (US).

GTD graphic tablet display.

GTDI *Guidelines for international trade data interchange, GEC*.

GTDPL generalized top down parsing language.

GTEPS General Telephone and Electronics Data Services.

GT/F General Telephone Company of Florida. Common carrier (US).

GTIS Gloucestershire technical information service (UK).

GTP graphics transform package.

GUIDE Guidance for Users of Integrated Data Processing Equipment. User Association (Switzerland).

GULP general utility library program.

H

H Henry. Unit of inductance.

HA 1. half adder. 2. *History abstracts*. Database, *ABC-Clio*. 3. home address.

HAB home address block.

HABS *Human relations area files automated bibliographic system*. Social science database, *HRAF*.

HADIS Huddersfield and district information service (UK).

HAI Holland Automation International. Software retailer.

HAIC hetero-atom in context. Indexing system, *CA*.

HAIT hash algorithm information table.

HAL 1. Harwell automated loans. Library circulation system, *UKAEA*. 2. hash algorithm library. 3. highly automated logic.

HALDIS Halifax and district information service (UK).

HALSIM hardware logic simulator.

HAM hierarchical access method.

HAMT human-aided machine translation.

HAPPE Honeywell associative parallel processing ensemble.

HAPUB high speed arithmetic processing unit board.

HAR home address register.

HARP Hitachi arithmetic processor.

HASL Hertfordshire Association of Special Libraries (UK).

HASP Houston automatic spooling processor. *IBM* equipment operating system.

HASQ hardware-assisted software queue.

HATREMS *Hazardous and trace emission system*. Databank, *EPA*.

HATRICS Hampshire technical research industrial and commercial service (UK).

HAYSTAQ have you stored answers to questions? National Bureau of Standards (US).

HAU 1. horizontal arithmetic unit. 2. hybrid arithmetic unit.

HAZFILE *Hazards file*. Online databank on emergency response, National Chemical Emergency Centre (UK).

HB hexadecimal to binary.

HBEN high byte enable.

HBR-online *Harvard Business Review – online*. John Wiley & Son (US).

HBZ Hochschulbibliothekszentrum. *ILL* and automated library system operator, Cologne (FRG).

HC 1. hard copy. 2. hardware capability. 3. headquarters city. Searchable field, Dialog *IRS*. 4. heated coil, telecommunications. 5. host computer. 6. hybrid computer.

HCC hardware capability code. Searchable field, Dialog *IRS*.

HCF host command facility.

HCG hardware character generator.

HCI host computer interface.

HCL Harold Cohen Library, University of Liverpool (UK).

HCP host communications processor.

HCR hardware check routine.

HCS host composition system. System to interface machine readable text, data and typesetting, Infograph Ltd.

HCW *Home Computing Weekly* (UK).

HD 1. half duplex, telecommunications. 2. hierarchical direct. 3. high density.

HDA head (and) disc assembly.

HDAM hierarchical direct access method.

HDAS hybrid data acquisition system.

HDB high density bipolar.

HDB$_3$ high density binary 3. *PCM* specification.

HDDR high density digital recording.

HDDS high density data system.

HDF high density flexible.

HDI head-disc interference, also head crash. Failure.

HDLC high level data link control. Protocol.

HDMR high density multi-track recording.

HDOS hard disc operating system.

HDR high density recording.

HDS 1. Hermes data system, Hermes Precisa International. 2. hybrid development system.

HDTV high definition television.

HDU hard disc unit.

HDX half duplex. Telecommunications mode.

HE 1. heading note. Searchable field, *SDC*. 2. historical period ending date. Searchable field, Dialog *IRS*.

HEALS Honeywell error analysis and logging system.

HEBIS Hessisches Bibliothekssystem. Automated cataloguing system, Hessen (FRG).

HECSAGON Horowitz-Eastman-Crane symbol array governed by orthodox notation.

HEEP 1. *Health effects of environmental pollution*. Database, *NLM/BIOSIS*. 2. Highway engineering exchange program. Planned computer program exchange (US).

HEF *Heated effluents*. Database, Cornell University (US).

HELP highly extendable language processor.

HELPIS Higher education learning programmes information service. Council for Educational Technology (UK).

HEMTs high electron mobility transistors.

HEP 1. *High energy physics*. Database, *DESY*. 2. homogenous element processor, Denelcor.

HEPI *HEP* index. *DESY*.

HERMAN *Hierarchical environmental retrieval for management access and networking*. Database on biology, Biological Information Service (US).

HERMES heuristic mechanized documentation information system (Romania).

HERS hardware error recovery system, Sperry Univac.

HERTIS Hertfordshire technical library and information service (UK).

HES house exchange system, telecommunications.

HEX hexadecimal. Numbering system with base 16.

HEXFET hexagonal field effect transistor.

HF high frequency. Approx. 10^7 Hz.

HFAS Honeywell file access system.

HFBC *High frequency broadcasting schedule*. Databank, *ITU*.

HFDF high frequency distribution frame, telecommunications.

HFPS high frequency phase shifter, telecommunications.

HI highest grant. Searchable field, Dialog *IRS*.

HIC hybrid integrated circuits.

HICLASS hierarchical classification. Library science.

HICS hierarchical information control system, Fujitsu (Japan).

HID *Housing industry dynamics*. Originator and databank (US).

HIDAM hierarchical indexed direct access method.

HI FI high fidelity, audio reproduction.

HIFT hardware implemented fault tolerance.

HIM hardware interface module.

HIP host information processor.

HIS 1. homogeneous information sets. 2. Honeywell Information Systems.

HISAM hierarchical indexed sequential access method. Information storage method.

HISARS *Hydrological storage and retrieval information system*. Databank, North Carolina State University (US).

HISDAM hierarchical indexed sequential direct access method.

HISTLINE *History of medicine online*. Database, *NLM*.

HIT 1. high isolation transformer. 2. Hitachi Limited. Computer manufacturer (Japan).

HITAC Hitachi Computer Services.

HKB *Hepatitis knowledge base*. Expert system, *NLM*.

HL 1. high level. 2. host language.

HLAF *high level arithmetic function*.

HLAIS high level analog input system.

HLDTL high level data transistor logic.

HLI host language interface.

HLL high level language.

HLML high level microprogramming language.

HLPI high level programming interface.

HLQL high level query language.

HLR high level representation.

HLS high level scheduler.

HLSE high level single ended.

HLU house logic unit.

HMI hardware monitor interface.

HMM hardware multiply module.

HMO hardware microcode optimizer.

HMOS high speed metal oxide semiconductor. Type of *ROM*.

HMOS-E *HMOS* – erasable. Type of *PROM*.

HMR hybrid modular redundancy.

HMSO Her Majesty's Stationery Office, Government publishing agency (UK).

HN 1. hardware capability name. 2. headquarters name. Searchable fields, Dialog *IRS*.

HNA 1. hierarchical network architecture. 2. Hitachi network architecture.

HNIL high noise immunity logic.

HOARS hands-on annotated recorded search. Technique for teaching online searching.

HOBITS Haifa online bibliographic text system. Software, University of Haifa (Israel).

HOF head of form.

HOL high order language.

HOLUG Houston On Line Users Group (US).

HOP hybrid operating program.

HOQ 1. *Hansard oral questions*. Full text database of oral parliamentary questions and answers, *QL* (Canada). 2. home office quote.

Horitz horizontal.

HOS high order software.

HP 1. Hewlett Packard (US). 2. historical period. Searchable field, Dialog *IRS*. 3. host processor.

HPA heuristic path algorithm.

HPCA high performance communications adaptor.

HPF high pass filter.

HPIB Hewlett Packard interface bus.

HPRU Handicapped Persons Research Unit (UK).

HPT hear per track, magnetic storage.

HQE *Hansard questions écrit*. Full text database of written parliamentary questions and answers, *QL* (Canada). Also *HWQ*.

HQO *Hansard questions orales*. Full text database of oral parliamentary questions and answers, *QL* (Canada).

HR 1. high reduction, of microforms. 2. hit ratio, *IR*. 3. holding register. 4. hour.

HRAF Human Relations Area Files. Originates *HABS* database (US).

HRC hypothetical reference circuit, telecommunications.

HRDP hypothetical reference digital path, telecommunications.

HRIS 1. *Highway research information service*. Database, US Department of Transportation. 2. House of Representatives information system (US).

HRM hardware read-in mode.

HRNES host remote node entry system.

HRP hypergroup reference pilot, carrier transmission, telecommunications.

HRS host resident software.

HRSS host resident software system.

HRX hypothetical reference connection, telecommunications.

HS 1. headquarters state. Searchable field, Dialog *IRS*. 2. hierarchically structured, indexing language. 3. historical period starting date. Searchable field, Dialog *IRS*.

HSAM hierarchical sequential access method.

HSB high speed buffer.

HSBA high speed bus adaptor.

HSC high speed concentrator.

HSD high speed data.

HS-DA high speed data acquisition.

HSDB high speed data buffer.

HSEL high speed selector channel.

HSELINE *Health and Safety Executive online*. Database, Health and Safety Executive (UK).

HSI *HERTIS* subject index.

HSL *Highway safety literature*. Database, *NHTSA*.

HSLA high speed line adaptor.

HSM 1. hierarchical storage manager. 2. high speed memory.

HSP high speed printer.

HSR high speed reader.

HSRIOP high speed *RAD* input/output processor, Xerox.

HSRO high speed repetitive operation.

HSS 1. hierarchy service system, Toshiba. 2. high speed storage.

HT 1. half tone, printing. 2. head per track. 3. high tension. 4. horizontal tabulate.

HTB hexadecimal to binary.

HTC hybrid technology computer.

HTE hypergroup translating equipment.

HTFS *Heat transfer and fluid flow service*. Nucleonics database, *UKAEA*.

HTL high threshold logic.

HTS 1. head track and selector. 2. host to satellite.

HUG Honeywell Users Group.

HULTIS Hull technical interloan scheme (UK).

HV hardware virtualizer.

HVG high voltage generator.

HVR 1. hardware vector to raster. 2. home video recorder.

HVTS high volume time sharing.

HW hard wired.

HWI hardware interpreter.

HWIM hear what I mean. Speech recognition system, *ARPA*.

HWQ *Hansard written questions*. Full text database of written parliamentary questions, *QL* (Canada).

HYB hybrid systems, computing or telecommunications.

HYCOTRAN hybrid computer translator.

Hz 1. headquarters zip code. Searchable field, Dialog *IRS*. 2. hertz. Unit of frequency.

I

I 1. current. 2. input. 3. instruction.

I (field) information field, also data field.

IA 1. image analysis. 2. indirect address. 3. integrated adaptor. 4. interchange address. 5. international alphabet.

IAA 1. Institute of Administrative Accounting and Data Processing Limited (UK). 2. *International aerospace abstracts*. Database on aerospace science and technology, *NASA*.

IAALD International Association of Agricultural Librarians and Documentalists.

IAB 1. Institut für Arbeitsmarkt und Berufsforschung der Bundesantalt für Arbeit. Originator, operator and database on economics and the labour market (GDR). 2. interrupt address to bus.

IABC International Advisory Committee on Bibliography, *UNESCO*.

IAC 1. Information Access Corporation. Originator of *MI* and *NNI* databases (US). 2. information analysis center. Projected subject-specific centres (US). 3. intelligent asynchronous controller. Computer terminal connector. 4. interactive array computer. 5. international algebraic compiler. 6. International Association for Cybernetics.

IACBDT International Advisory Committee on Bibliography, Documentation and Terminology, *UNESCO*.

IACP International Association of Computer Programmers.

IAD 1. initial address designator. 2. initiation area discriminator. Type of *CRT*.

IADIC integration analog to digital converter.

IADIS Irish Association for Documentation and Information Services.

IAEA International Atomic Energy Agency. Database originator and operator (Austria).

IAF interactive facility, *CDC*.

IA-5 international alphabet 5. 7-bit communications code.

IAG 1. *IFIP* Automatic Data Processing Group. 2. instruction address register.

IAIS *Industrial aerodynamics information service*. Database (UK).

IAL 1. international algebraic language, replaced by *ALGOL*. 2. international algorithmic language.

IALE instrumented architectural level emulation.

IALINE *Industries algro-alimentaires on-line*. Database, *CDIUPA*.

IAM 1. initial address message, telecommunications. 2. interactive algebraic manipulation. 3. intermediate access memory.

IAMACS International Association for Mathematics and Computers in Simulation.

IAMC Institute for Advancement of Medical Communication (US).

IAMCR International Association for Mass Communication Research.

IAML International Association of Music Libraries.

IAMS integrated academic information management system, *KRI*.

I & A indexing and abstracting.

I and D information and documentation.

IA-1 image array processor.

IAP 1. image array processor. 2. internal array processor.

IAR 1. instruction address register. 2. interrupt address register.

IARD Information Analysis and Retrieval Division, American Institute of Physics.

IAS 1. immediate access storage. 2. interactive applications supervisor. 3. interactive applications system. 4. International Applied Systems (US).

IASC International Association for Statistical Computing.

IASLIC Indian Association for Special Libraries and Information Centres.

IASR interruption address storage register.

IASSIST International Association for Social Science Information Services and Technology.

IAT Institute for Advanced Technology (US).

IATAE international accounting and traffic analysis equipment, telecommunications.

IATUL International Association of Technological University Libraries.

IA-2 international alphabet – 2. Standard telegraphy code.

IAURIF Institut d'Aménagement et d'Urbanisme de la Région de l'Ile de France. Database originator.

IB 1. identifier block. 2. input bus. 3. instruction bus. 4. interface bus. 5. internal bus.

IBA Independent Broadcasting Authority (UK).

IBAS Informationssystem Beliebiger Andwendungssystem. Automated library system (FRG).

IBBY International Board on Books for Young People (also IBBYP).

IBC integrated block controller.

IBDI International Bureau of Documentation and Information on Sport.

IBEDOC International Bureau of Education documentation and information system, *UNESCO*.

IBG inter block gap.

IBI Intergovernmental Bureau for Informatics. Concerned with transborder data flow.

IBI-ICC *IBI* – International Computation Centre.

IBIM Instituto de Información y Documentación en Biomedicina. Information broker (Spain).

IBIS 1. intelligent business information system. 2. international bank information system. 3. International Book Information Service (UK).

IBJ Data Industrial Bank of Japan database. Originator and databank on trade and economics (Japan).

IBM International Business Machines. Computer manufacturer (US).

IBMCUA *IBM* Computer Users' Association.

IBOL interactive business oriented language.

IB (I and II) *Information bank (I and II)*. Databases, *NYTIS*.

IBSR interactive bibliographic search and retrieval. *IR* technique.

IC 1. identification code. 2. information codes. Searchable field, *BLAISE*. 3. input circuit. 4. instruction cell. 5. instruction counter. 6. integrated circuit. 7. intelligent copier. 8. interface control. 9. International Centre for Registration of Serial Publications, *UNESCO*. 10. international patent classification. Searchable field, *SDC*.

ICA 1. integrated communications adaptor. 2. intercomputer adaptor. 3. International Communications Agency (US). 4. International Communications Association (US). 5. International Council on Archives.

ICAA International Civil Aviation Authority. Database originator (Canada).

ICADI Inter-American Center for Agricultural Documentation and Information, Inter-American Institute of Agricultural Science.

ICAM integrated communications access method.

ICAQUO *Inventory of contaminants in aquatic organisms*. Databank, *FAO*.

ICB 1. incoming calls barred, telecommunications. 2. internal common bus.

ICC 1. International Computing Centre, *UN*. 2. International Conference on Communications, *IEEE*. 3. international control centre, telecommunications. 4. invitational computer conference, for buyers. 5. issue category code. Searchable field, *SDC*.

ICCA Independent Computer Consultants Association (US).

ICCC International Council for Computer Communication (US).

ICCF interaction computing and control facility.

ICCP Institute for Certification of Computer Professionals.

ICCS inter computer communications system.

ICD International Congress for Data Processing.

ICDB integrated corporate data base.

ICDDB internal control description data base.

ICDF *Inorganic crystallographic data file.* Databank, *INKA.*

ICDL 1. integrated circuit description language. 2. internal control description language.

ICDR inward call detail recording.

ICE 1. incircuit emulator. 2. input checking equipment. 3. Istituto Nazionale per il Commercio con l'Estero. Originator and databank on business and economics (Italy). 4. integrated communications environment. Computer architecture, *CTL.*

ICECAN Iceland-Canada cable.

ICF 1. integrated control facility. 2. interactive communications feature.

ICG interactive computer graphics.

ICI intelligent communication interface.

ICIC International Copyright Information Centre, *UNESCO.*

ICIE *Infogrow communications information exchange.* Database directory, *AUSINET.*

ICIP International Conference on Information Processing.

ICIREPAT International Cooperation in Information Retrieval among Examining Patent Offices. Documentation and information retrieval cooperation promotional body. Also *CICREPATO.*

ICIST Institut Canadien de l'Information Scientifique et Technique. Also *CISTI.*

ICL 1. incoming line. 2. intercommunication logic. 3. intercomputer communication logic. 4. International Computers Ltd. Computer manufacturer (UK). 5. interpretive coding language.

ICM instruction control memory.

ICN integrated computer network.

ICO International Coffee Organisation. Produces *Coffeeline* database.

ICOGRADA International Council of Graphic Design Associations.

ICONCLASS iconography classification (Netherlands).

ICOS interactive *COBOL* operating system.

ICP 1. initial connection protocol. 2. international classification of patents, Council of Europe.

ICPDATA *International commodity production data.* Databank, United Nations Statistical Office.

ICPSR Inter-university Consortium for Political and Social Research. Data archive (US).

ICR 1. indirect control register. 2. input control register. 3. International Council for Reprography. 4. interrupt control register.

ICRDB *International cancer research databank.*

ICRH Institute for Computer Research in the Humanities.

ICRS *Index chemicals registry system.* Databank, *ISI.*

ICS 1. input control subsystem. 2. integrated communications system. 3. interactive communications software. 4. *Interlinked computerized system* (for food and agricultural statistics). Databank, *FAO.* 5. interpretive computer simulation.

ICSC 1. Interim Commission on Satellite Communication. 2. International Communications Systems Consultants. Originator, operator and databank on rates for leased circuits (UK). 3. Irvine Computer Sciences Corporation (US).

ICSMP interactive continuous systems modelling program.

ICSSD International Committee for Social Sciences Documentation and Information. Also *ICSSDI.*

ICSSDI see *ICSSD.*

ICSTI International Centre for Scientific and Technical Information (USSR).

ICSU International Council of Scientific Unions.

ICSU-AB *ICSU* Abstracting Board.

ICT 1. image creation terminal. Graphics system, Compuvision. 2. Institute of Circuit Technology Ltd (UK). 3. International Computers and Tabulators Ltd, now *ICL*.

ICU 1. instruction control unit. 2. integrated control unit. 3. interface control unit.

ICW 1. initial condition word. 2. interface control word.

I-cycle instruction cycle. Computing.

ICYT Instituto de Información y Documentación en Ciencia y Technologia. Database originator and host (Spain).

ID 1. identification. 2. identifier. User password to database. 3. identifier(s). Searchable field, *ESA-IRS* and Dialog. 4. insulation displacement. 5. instruction data. 6. intelligent digitizer.

IDA 1. integrated digital access, to *ISDN*. 2. integrated disc adaptor. 3. intelligent data access. 4. interactive debugging aid.

IDAM indexed direct access method.

IDAS information displays automatic drafting system.

IDB input data buffer.

IDBMS integrated *DBMS*.

IDC 1. integrated disc control. 2. Interactive Data Corporation. Division of Chase Manhattan Bank, hosts databases and databanks (US). 3. internal data channel. 4. Internationale Dokumentationgesellschaft für Chemie. Chemical information storage and retrieval development body (FRG, Austria, Netherlands).

IDCAS Industrial Development Centre for Arab States. Databank originator and operator (Egypt).

IDCC integrated data communications controller.

IDCHEC Intergovernmental Documentation Centre on Housing and Environment. Originator, operator and its databank (France).

IDCMA Independent Data Communications Manufacturers Association.

IDCS international digital channel service.

IDD 1. integrated data dictionary. 2. international direct dialling (UK).

IDDD international direct distance dialling (US).

IDDS international digital data service. Data transmission service.

IDE interactive data entry.

IDEA interactive data entry access, Data General.

IDEN interactive data entry network, Xerox.

IDEP International Data Exchange Program (US).

IDES Incoterm data entry software, Incoterm.

IDF 1. image description file. 2. integrated data file. 3. intermediate distribution frame, telecommunications. 4. inverse document frequency, *IR* weighting technique.

IDI intelligent dual interface.

IDIOT instrumentation digital online transcriber.

IDIS 1. Institut für Dokumentation und Information über Socialmedizin und Offentliches Gesundheitswesen. Originator and database on public health (FRG). 2. Iowa Drug Information Service. Database originator and information system, University of Iowa (US).

IDL 1. information description language. 2. instruction definition language.

IDM Interactive Data Machines (UK).

IDMAS interactive database manipulator and summarizer.

IDMH input destination message handler.

IDMS integrated database management system.

IDN 1. integrated digital network. 2. intelligent data network.

IDOS 1. interactive disc operating system. 2. interrupt disc operating system.

IDP 1. industrial data processing. 2. input data processor. 3. Institute of Data Processing (UK). 4. integrated data processing. 5. interdigit pause, in processing.

IDPM Institute of Data Processing Management (UK).

IDPS interactive direct processing system, *NCR*.

IDPS/LF IDPS/large file, *NCR*.

IDR 1. industrial data reduction.
2. information dissemination and retrieval, Reuters.

IDS 1. idle signal unit, telecommunications.
2. information display system, *HP*.
3. integrated data storage. 4. intelligent display system, Xerox. 5. interactive design software, *HP*. 6. interactive display system.

I-D-S integrated data store, Honeywell.

IDST information and documentation on science and technology.

IDT intelligent data terminal.

IDW Institut für Dokumentationswesen (FRG).

IDX intelligent digital exchange, Plessey.

IE interrupt enable.

IEA International Energy Agency. Originator and database (UK).

IEC 1. *Information exchange centre.* Databank on current research in Canadian universities, *CISTI*. 2. International Electrotechnical Commission. International standards body.

IEE Institution of Electrical Engineers (UK).

IEEE Institute of Electrical and Electronics Engineering (US).

IEF instruction execution function.

IEO integrated electronic office.

IEPRC International Electronic Publishing Research Centre (UK).

IERE Institution of Electronic and Radio Engineers (UK).

IES information exchange system, *ESPIRIT*.

IETEJ Institute of Electronics & Telecommunications Engineers of Japan.

IEV international electrotechnical vocabulary.

IF 1. instruction field. 2. interface.
3. intermediate frequency. 4. inverted file.

IFA 1. information flow analysis. 2. integrated file adaptor.

IFAC International Federation for Automatic Control.

IFACE interface element, computing.

IFAM inverted file access method.

IFCS International Federation for Computer Sciences.

IFD International Federation for Documentation, also *FID*.

IFEP integrated front end processor, data communications.

IFGL initial file generation language.

IFI Information for Industry Inc. (US).

IFILE interface file.

IFIP International Federation for Information Processing.

IFIPS International Federation of Information Processing Societies.

IFIS 1. international financial intelligence service (proposed), Financial Times and *ITT*. 2. *International food information service.* Current awareness service, Commonwealth Bureau of Dairy Science and Technology (UK).

IFLA International Federation of Library Associations.

IFM interactive file manager.

IFO International Fortran Organization (US).

IFORS International Federation of Operational Research Societies.

IFP 1. Institut Francais du Pétrole. Information broker, host and databank originator (France). 2. integrated file processor.

IFR 1. interface register. 2. internal function register.

IFRB International Frequency Registration Board. Radio frequency registration body of *ITU*.

IFS 1. interactive file sharing. 2. interchange file separator. 3. *International financial statistics.* Databank, *IMF*.

IFSEA International Federation of Scientific Editors Associations.

IFU instruction fetch unit.

IGC Institute for Graphic Communication (US).

IGFET insulated gate field effect transistor.

IGL interactive graphics language.

IGS 1. information group separator. 2. integrated graphic system.

IGT intelligent graphics terminal, Tektronix.

IH interrupt handler.

IHS Information Handling Services. Owns *BRS* host (US).

II interrupt inhibit.

IIA Information Industry Association (US).

IIASA International Institute for Applied Systems Analysis (Austria).

IIC International Institute of Communications.

IICMFA Integrated Information Centre of the Ministry of Foreign Affairs (Saudi Arabia).

IIDA indivisualized instruction for data access. Education package, Drexel University and Franklin Institute (US).

IIL integrated injection logic, microprocessing.

I Inf Sci Institute of Information Scientists (UK).

IIOP integrated input/output processor.

IIR *International inventors registry*. Database, Control Data Technotec Inc. (US).

IIRS Institute for Industrial Research and Standards. *ESA* broker (Ireland).

IIS 1. Institute of Information Scientists (UK). 2. interactive instructional system.

IIT industrial information transfer.

IIU instruction input unit.

IJCAI International Joint Conference on Artificial Intelligence.

IJS interactive job submission.

IKBS intelligent knowledge-based system, using *AI*.

IL 1. instruction list. 2. intermediate language.

ILA intelligent line adaptor.

ILACS integrated library administration and cataloguing system. Computerized library system, Unilever (Netherlands).

ILB initial load block.

ILC 1. instruction length code. 2. instruction location counter.

ILD *International labour documentation*. Database, *ILO*.

ILE interfacing latching element.

ILEM inter-library electronic mail.

ILIC International Library Information Center. University of Pittsburgh Graduate School of Library and Information Science (US).

ILL inter-library loan.

ILLINET Illinois library network. Library cooperative (US).

ILM 1. information logic machine. 2. intermediate language machine.

ILO 1. individual load operation. 2. International Labour Office. Part of UN, database originator and operator.

ILP 1. intermediate language processor. 2. intermediate language program.

ILR 1. Institute of Library Research, University of California (US). 2. instruction location register.

ILS 1. integrated library system, *NLM*. 2. integrated library system, Easy Data Systems (Canada). 3. international line selector.

ILTMS international leased telegraph message switching. Network service, British Telecom.

IM 1. instruction memory. 2. integrated modem. 3. intermodulation, telecommunications. 4. interrupt mask.

IMA 1. input message acknowledgement. 2. invalid memory address.

IMAC Illinois microfilm automated cataloging. Illinois State Library (US).

IMAG Instituut voor Mechanistie Arbeid en Gebouwen (Institute of agricultural engineering). Originator and its agricultural databank (Netherlands).

IMB inter-module bus.

IMC 1. integrated multiplexer channel. 2. intelligent matrix control. 3. interactive module controller. 4. International Information Management Congress, formerly International Micrographic Congress.

IMD Institüt für Maschinelle Dokumentation. Information broker, host and database originator (Austria).

IMDS International Microform Distribution Service.

IME 1. Industria Machine Electroniche. Computer manufacturer (Italy). 2. International Microcomputer Exhibition.

IMF 1. Institut de Mécanique des Fluides. Originator and database on fluid mechanics (France). 2. International Monetary Fund. Databank originator.

IMI 1. intermediate machine instruction. 2. International Market Intelligence. Databank originator (Norway).

IMIA International Medical Informatics Association. Special interest group of *IFIP*.

IMIS integrated management information system.

IML 1. information manipulation language. 2. initial machine load. 3. interactive maintenance language, Denelcor. 4. intermediary music language. Input code for music notation.

IMM intelligent memory manager.

IMOS interactive multiprogramming operating system, *NCR*.

IMP 1. integrated message processor. 2. integrated microprocessor. 3. intelligent message processor, Delta Data Systems. 4. interface message processor. 5. intrinsic multiprocessing.

IMPACT inventory management program and control techniques, *IBM*.

IMPCON inventory management and production control. Software, *BLSL*.

IMPL 1. implementation language. 2. initial microprogram load.

IMPS 1. integrated modular panel system. 2. interface message processors, computing. 3. International Micro Programmers' Society.

IMR 1. intelligent machine research. 2. interrupt mask register.

IMRADS information management retrieval and dissemination system.

IMS 1. information management system, *IBM*. 2. Interactive Market Systems. Host (US). 3. International Musicological Society. Co-originates *RILM*.

IMSI information management system interface.

IMSP integrated mass storage processor.

IMS/VS information management system/ virtual storage.

IMT integrated microimage terminal, *KAR* system, Kodak.

IMU 1. increment memory unit. 2. instruction memory unit.

IMX Inquiry Message Exchange (US).

IN 1. inventors. Searchable field, Pergamon-Infoline. 2. investigator name. Searchable field, Dialog and *SDC*.

INA 1. Institut National de la Communication Audiovisuelle (France). 2. integrated network architecture.

INC incorporated.

INCIRS international communication information retrieval system, University of Florida (US).

INCL include.

INCLUDE implementing new concepts of the library for urban disadvantaged ethnics. Cleveland Public Library (US).

INCOLSA Indiana Cooperative Library Services Authority (US).

INCOMEX International Computer Exhibition.

101

INDIS industrial information system, *UN* Industrial Development Organization.

INFAS Institut für Angewandte Socialwissenschaft. Host offering election result data (FRG).

INFCO Information Committee of the International Standards Organization.

INFIRS *Inverted file information retrieval system*. UK Chemical Information Service.

INFO information network and file organization.

INFOHOST *Information on hosts*. Database guide to German hosts, *GID*.

INFOL information oriented language, for *IR*.

INFOLAC Information for Latin American Countries project, *UN*.

INFOR information oriented language. Computer programming language.

INFORM 1. microcomputer-based information dissemination system. PL Systems Inc. (US). 2. see *ABI/INFORM*.

INFORMALUX Information Luxembourg. Host service of the Centre d'Énergie Informatique (Luxembourg).

INFOS *Informationszentrum für Schnittwerte*. Database on machinery, *RWTH*.

INFOTERM International Information Centre for Terminology (Austria).

INFRAL information retrieval automatic language. Adaptation of *COBOL* and *ALGOL*.

INFROSS Investigation into information requirements of social sciences. 1970s study (UK).

INGA interactive graphics analysis.

INGRES interactive graphic and retrieval system.

INIBON AN ASSR Institut Naucnoi Informacii i Fundamentalnaja Biblioteka po Obscestvennym Naukam, Akademii Nauk SSR. Institute of Scientific Information and Central Library on Social Science, Academy of Sciences (USSR).

INID Institut National de Informare si Documentare. National Institute for Information and Documentation, Science and Technology (Romania).

INIS Internation Nuclear Information Service. Information service on nuclear science and technology, International Atomic Energy Authority.

INKA Informationssystem Karlsruhe. Host (FRG).

INKA-CONF *INKA conference announcement*. Databank, *INKA*.

INKA-CORP *INKA corporations*. Databank, *INKA*.

INKA-DATACOMP *INKA data compilations*. Databank on physics, *INKA*.

INKA-MATH *INKA mathematics*. Database on mathematics literature, *INKA*.

INKA-NUCLEAR *INKA nuclear science and technology*. Database (includes *INIS*), *INKA*.

INKA-PHYS *INKA physics*. Database, *INKA*.

INMAP Independent Microelectronics Applications. Advice and consultancy company established jointly by Edinburgh and Heriot Watt universities (UK).

INMARSAT International Maritime Satellite. Satellite communications organization.

INP 1. integrated network processor. 2. intelligent network processor, *HP*.

INPADOC *INKA patent documentation*. Patents database, *INKA*.

INPI Institut National de la Propriété Industrielle. Patent Office (France).

INPI 1 *INPI database 1*. Database on French patents, *INPI*.

INPI 2 *INPI database 2*. Database on European patents, *INPI*.

INRA Institut National de la Recherche Agronomique. Database originator and operator (France).

INRIA Institut National de Recherche en Information et Automatique. Research centre and database originator (France).

INS integrated networking system. *IBM* switching system.

INSAR instruction address register.

INSAT India satellite, for telecommunications.

INSCOPE information system for coffee and other product economics, International Coffee Organisation.

INS Data Base *United States international air travel statistics data base.* Databank, I. P. Sharp Associates (Canada).

INSDOC Indian National Scientific Documentation Centre.

INSEE Institut National de la Statistique et des Études Économiques. Databank originator and operator (France).

INSERM Institut National de la Santé et de la Recherche Médicale. Host and databank originator (France).

INSIS inter-institutional integrated services information system. *IT* implementation project, *CEC*.

INSPEC *Information service: physics, electrical and electronics, and computers and control.* Information service, *IEE* (UK).

INSPEL *International newsletter of special libraries, IFLA.*

INSTAB Information service on toxicology and biodegradability, Water Pollution Research Laboratory (UK).

INSTARS information storage and retrieval systems.

INTA Instituto Nacional de Técnica Aeroespacial 'Esteban Terradas'. Information broker (Spain).

INTAG International Advisory Group on Technology Management. Information broker and consultancy.

INTD Institut National des Techniques de la Documentation (France).

INTEBRID monolithic integrated and hybrid circuitry, *CDC*.

INTEC Interface Technology (UK).

INTELPOST international telecommunications post. Facsimile transmission service, *INTELSAT*.

INTELSAT International Telecommunications Satellite Consortium.

INTERMARC international *MARC*.

INTFU interface unit.

INTGEN interpreter generator.

INTIME interactive textual information management experiment, *IBM*.

INTIPS integrated information processing system.

INTRAN input translator. *IBM* system.

INTREDIS *International tree disease register.* Database, US Forest Service/*NAL*.

INTREX information transfer experiment. *MIT* (US).

INVIDO système d'informations visuelles à domicile. System for study of user reaction to videotex (Canada).

INWATS inward wide area telephone service.

INWG International Network Working Group. Network standards and protocols working group, *IFIP*.

IO interpretive operation.

I/O input/output, computing.

IOA input/output adaptor.

IOAU input/output access unit.

IOB 1. input/output buffer. 2. Inter-Organization Board for Information Systems and Related Activities, *UN*.

IOBFR input/output buffer, computing.

IOBS input/output buffering system, computing.

IOC 1. input/output channel. 2. input/output connector. 3. input/output control. 4. input/output controller. 5. Intergovernmental Oceanographic Commission. Originator and operator of databases and databanks, *UNESCO*.

I/OC input/output controller.

IOCTR input/output controller, computing.

IOCS input/output control system.

IOD 1. Information on Demand. Document fulfilment agency (US). 2. input/output device. 3. *Indeks informasjon og dokumentasjon*. Database, *NSI*.

IOF 1. interactive operations facility. 2. input/output front end.

IOIH input/output interrupt handler.

IOLA 1. input/output line adaptor. 2. input/output link adaptor.

IOLC input/output link controller.

IOLIM International Online Information Meeting (UK).

IOM input/output multiplexer.

IOMP input/output microprocessor.

IOOP input/output operation, computing.

IOP input/output processor.

IOPG input/output processor group.

IOPKG input/output package. *IBM* system.

IOPS input/output programming system.

IOQ input/output queue.

IOR input/output register.

IORB input/output record block.

IOREQ input/output request.

IOS 1. input/output selector. 2. input/output subsystem. 3. input/output supervisor. 4. input/output system, General Automation. 5. interactive operating system.

IOSS input/output subsystem.

IOT 1. input/output and transfer. 2. input/output transfer. 3. input/output trunk.

IOTA information overload testing apparatus.

IOTG Input/Output Task Group, *CODASYL*.

IP 1. impact printer. 2. information provider, usually of viewdata. 3. instruction pointer. 4. interface processor.

IPA 1. information processing architecture. *ICL* software 2. Information Processing Association (Israel). 3. integrated peripheral adaptor. 4. integrated printer adaptor. 5. *International pharmaceutical abstracts*. Database, American Society of Hospital Pharmacists. 6. International Publishers Association.

IPB 1. integrated processor board. 2. interprocessor buffer.

IPC 1. independent control point, computing. 2. industrial process control. 3. information processing centre. 4. information processing code. 5. integrated peripheral channel. 6. integrated printer adaptor. 7. international patent classification, Council of Europe. 8. inter-process communication. 9. inter-process controller. 10. inter-process coupler.

IPCF 1. inter-process communication facility. 2. inter-program communication facility, Prime Computers.

IPCS interactive problem control system.

IPDC International Programme for the Development of Communication, *UNESCO*.

IPE information processing equipment.

IPG 1. Information Policy Group, *OECD*. 2. *INPADOC patent gazette, COM SDI, INPADOC*. 3. interactive presentation graphics, *IBM*.

IPI 1. individually presented instruction. 2. intelligent printer interface.

IPL 1. information processing language. 2. initial program load(ing). 3. interrupt priority level.

IPLC international private leased circuits, British Telecom International.

IPL-V information processing language – five. High level processing language.

IPM 1. impulses per minute. 2. interruptions per minute.

IPN 1. initial processing number. 2. *International publishing newsletter* (US).

IP/OP input/output interface element, computing.

IPPF instruction preprocessing function.

IPRA International Peace Research Association. Database originator (Netherlands).

IPS 1. impulses per second. 2. inches per second. 3. installation performance

specification. 4. instructions per second.
5. intelligent printing system.
6. interruptions per second.

IPSA I. P. Sharp Associates. Database producer and marketer.

IPSB interprocessor signal bus.

IPSE integrated programming support environment.

IPSJ Information Processing Society of Japan.

IPSO interface peripheral standard Olivetti, Olivetti.

IPSOC Information Processing Society of Canada.

IPSS international packet switched service. Telecommunications system.

IPSSB Information Processing Systems Standards Board (US).

IPTC International Press Telecommunications Council.

IPTM interval pulse time modulation, telecommunications.

IPU 1. instruction processing unit. 2. interprocessor unit.

IQF interactive query facility.

IQL 1. information query language. Command language. 2. interactive query language, *DEC*.

IQM input queue manager, computing.

IQMH input queue message handler.

IQPP interactive query pre-processor, *IR*.

IR 1. index register. 2. information retrieval. 3. infrared. 4. instruction register. 5. interrupt register.

IRA information resource administration.

IRAM indexed random access memory.

IRANDOC Iranian Documentation Centre.

IRAS Information Retrieval Advisory Services Limited (UK).

IRB 1. Informationsverbundzentrum Raum und Bau. Database originator and operator (FRG). 2. interruption request block.

IRBEL *indexed references to biomedical engineering literature.* National Institute for Medical Research (UK).

IRC 1. information retrieval centre. 2. international record carrier, telecommunications. 3. International Reference Centre, for community water supply and sanitation, *UN*.

IRCIHE International Referral Centre for Information Handling Equipment. *UNISIST UNESCO* centre (Yugoslavia).

IRCOL Institute for Information Retrieval and Computational Linguistics, Bar Ilam University (Israel).

IRCS International Research Communications System. Electronic journal publisher (UK).

IRD International Resource Development. *IT* market research organization (US).

IRDS *International road documentation scheme.* Database, *OECD*.

IRE Institute of Radio Engineers (US).

IRED infrared emitting diode.

IRF input register full.

IRFA Institut de Recherches sur les Fruits et Agrumes. Originator, operator and its database (France).

IRFITS infrared fault isolation test system.

IRG inter-record gap, on magnetic tape.

IRGMA Information Retrieval Group of the Museums Association (UK).

IRH inductive recording head.

IRIA 1. Infrared Information Analysis Center, University of Michigan (US). 2. Institut de Recherche d'Informatique et d'Automatique. Research centre, host and database originator (France).

IRIS 1. *Instructional resources information system.* Database, Ohio State University (US). 2. international research information service, American Foundation for the Blind (US).

IRJE interactive remote job entry.

IRL 1. information retrieval language. 2. Information Retrieval Limited. Database originator (UK).

105

IRM information resource management.

IRMS information retrieval and management system, *IBM*.

IRN internal routing network.

IROS instant response order system. Teleordering system, Brodart (US).

IRPL *Index to religious periodical literature*. Database (US).

IRQ interrupt request.

IRR interrupt return register.

IRRD *International road research documentation*. Database on roads and traffic, *ESA-IRS*.

IRRL Information Retrieval Research Laboratory, University of Illinois (US).

IRS 1. Information Retrieval Service. Host, *ESA*. 2. information retrieval system. 3. interchange record separator. 4. *International referral system for sources of environmental information*. Database, *UNEP*.

IRSC inter-regional subject coverage scheme. Libraries cooperative scheme (UK).

IRSCS see *IRSC*.

IRSG Information Retrieval Specialist Group, British Computer Society.

IRT 1. Institut de Recherche des Transports. Originates Urbamet database (France). 2. Institute of Reprographic Technology (UK).

IRTU intelligent remote terminal unit.

IRTV information retrieval television. Tele-education project (US).

IRU indefeasible right of use. Guarantee for cable network subscriber.

IRV interrupt request vector.

IRW indirect reference word.

IRX interactive resource executive, *NCR*.

IS 1. information separation. 2. *ISSN*. Searchable field, *NLM*. 3. issue number. Searchable field, Dialog and *SDC*.

ISA 1. *Information science abstracts*. Database, Plenum Publishing Corporation (US). 2. Instrument Society of America. 3. interrupt storage area.

ISAD Information Science and Automation Division, *ALA*, now *LITA*.

ISAL information system access line.

ISAM 1. indexed sequential access method, *IBM*. 2. integrated switching and multiplexing.

ISAR information storage and retrieval.

ISB independent sideboard transmission, telecommunications.

ISBD international standard bibliographic description.

ISBD (CM) *ISBD* cartographic materials.

ISBD(G) *ISBD* general.

ISBD(M) *ISBD* monographs.

ISBD(NBM) *ISBD* non-book materials.

ISBD(S) *ISBD* serials.

ISBF interactive search of bibliographic files.

ISBL information system base language.

ISBN international standard book number.

ISBS integrated small business software.

ISC 1. Information Science Corporation. Host and database originator (US). 2. Information Society of Canada. 3. integrated storage control. 4. intelligent synchronous controller, computing. 5. Intelligent Systems Corporation (US). 6. international switching centre, telecommunications. 7. inter-system communication.

ISCED international standard classification of education, *UNESCO*.

ISD 1. information structure design. 2. initial selection done. 3. intermediate storage device. 4. international subscriber dialling, now *IDD*.

ISDG Information Science Discussion Group (UK).

ISDN 1. integrated services digital network. Extension of *IDN* to accommodate services such as fax, electronic mail. 2. international standard data network.

ISDOS information system design by optimization system. Computer produced computer programs.

ISDS 1. integrated ship design system. 2. integrated software development system. 3. international serials data system, *UNISIST*.

ISE 1. Institute for Software Engineering (US). 2. in system evaluator, National Semiconductor Company. 3. interrupt system enable. 4. inter system emulator.

ISF 1. individual store and forward. 2. information systems factory. Computer system for producing *IT* systems.

ISFD integrated software functional design.

ISFM indexed sequential file manager.

ISG inter subblock gap.

ISI 1. Institute for Scientific Information. Originates *SCI* and related databases, offers current awareness services (US). 2. internally specified index. 3. inter symbol interference.

ISIC international standard industrial classification, UN.

ISI/ISTP&B *ISI index to scientific and technical proceedings and books*. Database, *ISI*.

ISILT information science index language text, *CLW*.

ISIS internally switched interface system.

ISK 1. insert storage key. 2. instruction space key.

ISL 1. information search language. 2. information system language. 3. instructional systems language. 4. interactive simulation language. 5. inter-satellite link. 6. inter-system link.

ISLA International Survey Libraries Association, University of Connecticut (US).

ISM 1. information systems for management. 2. International Software Marketing (US).

ISMEC *Information service in mechanical engineering*. Database, *IEE* (UK) and *DCI* (US).

ISMH input source message handler.

ISMS image store management system.

ISO 1. individual system operation. 2. International Standards Organization.

ISOC Instituto de Información y Documentación en Ciencia Sociales. *ERIC* operator (Spain).

ISORID *International information system on research in documentation*. Database, *UNESCO*.

ISP 1. indexed sequential processor. 2. instruction set processor.

ISPW International Society for the Psychology of Writing (Italy).

ISR 1. image storage retrieval. 2. *Index to scientific reviews*. Current awareness service, *ISI*. 3. information storage and retrieval. 4. Innovative Systems Research. Experimented with communications systems on technology for the handicapped (US). 5. interrupt service routine.

ISRD information storage retrieval and dissemination.

ISRM information system resource manager.

ISS 1. information sharing system. 2. information storage system. 3. Information Systems Specialists Office, Library of Congress (US). 4. input subsystem. 5. integrated storage system. 6. intelligent support system.

ISSN international standard serial number.

ISSOC International Solid State Circuits Conference.

ISSP information system for policy planning.

ISSUE information system software update environment.

IST 1. information science and technology. 2. integrated system test.

ISTC Institute of Scientific and Technical Communicators (UK).

ISTIM interchange of scientific and technical information in machine language.

ISTP&B *Index to scientific and technical proceedings and books*. Database, *ISI*.

ISTR indexed sequential table retrieval.

ISU 1. initial signal unit, telecommunications. 2. instruction storage unit. 3. interface sharing unit. 4. interface switching unit.

IT 1. index terms. Searchable field, *NLM* and *SDC*. 2. information technology. 3. information theory. 4. input terminal. 5. input translator. 6. intelligent terminal. 7. internal translator, Carnegie Institute (US). 8. item transfer.

ITA 1. Institut du Transport Aérien. Originator and operator of database on international commercial aviation (France). 2. interface test adaptor. 3. International Tape Association. 4. international telegraph alphabet.

ITAL *Information technology for libraries*. Journal, *LITA*, formerly *JOLA*.

ITAP Information Technology Advisory Panel (UK).

ITAVS integrated testing analysis and verification system.

ITB 1. intermediate test block. 2. internal transfer bus.

ITBTP Institut Technique du Bâtiment et des Travaux Publics. Originator and database (France).

ITC 1. Information Technology Ltd (UK). 2. integrated terminal controller. 3. Interdata transaction controller, Perkin-Elmer. 4. International Teletraffic Congress, for *PTT* and telecommunications industry. 5. International Telemetering Conference. 6. International Trade Commission. Databank originator (US). 7. International Translation Centre. Originates *World transindex* databank.

ITCA International Typographical Composition Association.

ITCC International Technical Communication Conference, Society for Technical Communications (US).

ITD *Information trade directory*. International directory of information products and services, Gale Research Co. (US).

ITDM intelligent time division multiplexer.

ITE Institute of Telecommunications Engineers.

ITEC(S) information technology centre(s). Training centres (UK).

ITELIS *Irish Times Eurolex legal information service*. Database.

ITEMS Incoterm transaction entry management system, Incoterm.

ITF 1. Institut Textile de France. Operator and originator of *TITUS*. 2. integrated test facility. 3. interactive terminal facility.

I3L insoplanar integrated injection logic.

ITI interactive terminal interface.

ITIL International Tsunami Information Center, Hawaii (US).

ITIRC IBM Technical Information Retrieval Center (US).

ITL 1. intermediate text language. 2. intermediate transfer language.

ITM 1. indirect tag memory. 2. information transfer module, telecommunications.

ITN integrated teleprocessing network.

ITOS interactive terminal operating system, *CDC*.

ITP interactive terminal protocol.

ITPS 1. interactive teleprocessing system. 2. interactive text processing system.

ITR 1. interrupt control unit, computing. 2. isolation test routine.

ITS 1. intelligent terminal system. 2. interactive terminal service, Xerox. 3. interactive terminal support. 4. International Time-sharing Corporation. Telecommunications company and databank originator (US). 5. invitation to send, data communications.

ITSC 1. International Telecommunications Satellite Consortium (US). 2. international telephone service centre.

ITSU Information Technology Standards Unit, *DoI*.

ITT 1. Institute of Textile Technology. Databank originator (US). 2. International Telephone and Telegraph Corporation (US).

ITU 1. International Telecommunications Union. Telecommunications agency (UN). 2. International Typographical Union.

ITV 1. independent television (UK). 2. instructional television.

I²L integrated injection logic, *IC* technology.

IU 1. information unit. 2. input unit. 3. instruction unit. 4. interface unit.

IUL Information Utilization Laboratory, University of Pittsburgh (US).

IURP integrated unit record processor.

IUS 1. information unit separator. 2. interchange unit selector. 3. interchange unit separator.

IUTS inter-university transit system. Inter-library loan service (Canada).

IV investigator. Searchable field, Dialog *IRS*.

IVCE International Video and Communications Exhibition.

IVG interrupt vector generator.

IVIPA International Videotex Information Providers Association.

IVM 1. initial virtual memory. 2. interface virtual machine, computing.

IVP installation verification procedure.

IVT integrated video terminal.

IW index words. Searchable field, *SDC*.

IWB instruction word buffer.

IWIM Institut für Wissenschaftinformation in der Medizin. Institute for scientific information in medicine (GDR).

IWP International Word Processing Association (US-based).

IX 1. inter-exchange, telecommunications. 2. inverted index.

IXC inter-exchange control.

IXSD international telex subscriber dialling.

IZ Informationszentrum für Sozialwissenschaftliche Forschung. Originator and operator of social science database (FRG).

J

JA 1. journal announcement. Searchable field, Dialog *IRS*. 2. journal issue. Searchable field, *ESA-IRS*. 3. jump address.

JAF job accounting facility.

JAI job accounting interface.

JAMASS Japanese medical abstract scanning system, International Medical Information Center (Japan).

JAPATIC Japan Patent Information Center. Host (Japan).

JAR jump address register.

JASIS *Journal of the American Society for Information Science.*

JAT job accounting table.

JC 1. journal citation, 2. journal coden. Searchable fields, *SDC*. 3. journal title code. Searchable field, *NLM*.

JCB job control block.

JCC 1. job control card. 2. Joint Computer Conference (US). 3. Joint Consultative Committee, of information professional institutions (UK).

JCL 1. job command language. 2. job control language, computing.

JCN junction.

JCP job control program.

JCPDS Joint Committee on Power Diffraction Standards. Databank originator (US).

JCT job control table.

JCTND Jugoslavenski Centar za Tehniku i Naucno Dokumentaciju. Yugoslav centre for technical and scientific documentation.

JDL 1. job description library. 2. job descriptor language.

JEAN *JOSS*-based expression analyser for the nineteen hundred, *ICL*.

JECC Japan Electric Computer Corporation. Computer manufacturer (Japan).

JECL job entry control language.

JECS job entry control services, *IBM*.

JEDEC Joint Electron Device Engineering Council. Hardware and software standards body (US).

JEIDA Japanese Electronic Industry Development Association.

JEIPAC *JICST* electronic information processing automatic computer.

JEN Junta de Energia Nuclear. *INIS* operator (Spain).

JEPS job entry peripheral services, *IBM*.

JES 1. job entry subsystem, *IBM*. 2. job entry system.

JETS job executive and transport satellite, *NCR*.

JEUDEMO jeu de mots. Text handling package, University of Montreal (Canada).

JFCB job file control block.

JFET junction field effect transistor.

JFN job file number.

JG junction grammar. Machine translation term.

JIB job information block.

JICST Japan Information Center of Science and Technology. Database producer and host.

JIIM *Journal of information and image management*, formerly *Journal of micrographics.*

JIP joint input processing.

JIRC *Journal of information research communications* (UK).

JIS 1. Japanese industrial standard. 2. job information system. 3. job input system. 4. *Journal of information science, IIS.*

JK (flip flop) J and K input flip flop. Logic storage element.

JM *Journal of micrographics*, now *JIIM*.

JMSX job memory switch matrix.

JN 1. journal name. Searchable field, Dialog *IRS*. 2. source journal. Searchable field, *ESA-IRS*.

JNT Joint Network Team, of the Computer Board (UK).

JO 1. job order. 2. journal name announcement or citation. Searchable fields, Dialog *IRS*.

JOIS Japanese online information system. Bibliographic retrieval system, *JICST*.

JOIS II see *JOIS*.

JOL job organization language.

JOLA *Journal of library automation, ISAD*, now *ITAL*.

JOSS Johnniac open-shop system. Programming language for mathematics.

JOVIAL Jule's own version of international algorithmic language. Command and control language.

JP job processor.

JPA job pack area.

JPRA Japanese Phonograph Record Association.

JPU job processing unit.

JPW job processing word.

JRC Joint Research Centre of the *CEC*. Host.

JRS junction relay set, telecommunications.

JSCLC Joint Standing Committee on Library Cooperation (UK).

JSF job services file.

JSL job specification language.

JSLS Japan Society of Library Science.

JSRU Joint Speech Research Unit (UK).

JTE junction tandem exchange, telecommunications.

JTMP job transfer and manipulation protocol.

JTPS job and tape planning system.

JUG Joint Users Group. Computer user group (US).

JURIS 1. *Juristische Informationssystem*. Database on law, Federal Ministry of Justice (FRG). 2. *Department of Justice retrieval and inquiry system*. Databank, US Department of Justice.

JUSE Japanese Union of Scientists and Engineers. Databank originator.

JUSE-AESOPP *JUSE – an estimator of physical properties*. Databank, *JUSE*.

JVC Victor Company of Japan. Electronics and video manufacturer.

K

K 1. kilo (10^3). 2. 10^3 bits, computing.

KACHAPAG *Karlsruhe charged particle group*. Databank, *INKA*.

KAR Kodak automated retrieval. Microfilm office information system, Kodak.

KASC Knowledge Availability Systems Center, University of Pittsburgh (US).

KASS Kent automated serials system. Automated library system, Kent State University (US).

KATI Titelkatalogisierung von flaufnamen. Cataloguing software, University of Bielefeld (FRG).

KB 1. keyboard. 2. kilobit (10^3 bits). 3. kilobyte.

KBDSC keyboard and/or display controller, computing.

KBE *Key British enterprises*. Database, *D & B*, available on Pergamon Infoline.

KBENC keyboard encoder, computing.

KBPRC keyboard and printer controller, computing.

KBPS 1. kilobits per second. 2. kilobytes per second.

KBS kilobits per second.

KC kilocycle.

KCC keyboard common contact.

KCP keyboard-controlled phototypesetter.

KCS kilo characters per second.

KCU keyboard control unit.

KDD Kokusai Denshin Denwa Co. Ltd. Telecommunications company (Japan).

KDE keyboard data entry.

KDEM Kurzweil data entry machine. Optical character recognition equipment.

KDOS 1. key display operating system. 2. key to disc operating system.

KDP keyboard display and printer.

KDR keyboard data recorder.

KDS 1. key data station. 2. key to disc(ette) system.

KDT key data terminal.

KEIS Kentucky Economic Information Systems. Database producer.

KEP key entry processing.

KEYTECT keyword detection. *IR* program.

KF key field.

KFA Kernforschungsanlage julich. Databank originator and database operator (FRG).

K factor refers to bending of radio beams.

KFAS keyed file access system.

KGM key generator module.

KHz kilohertz (10^3 Hertz).

KICU keyboard interface control unit.

KIE Kirklees Information Exchange, formerly *HADIS* and *KIS*.

KIL keyed input language.

KINGMAP King's music analysis package. Program, King's College, University of London (UK).

KIOPI Kienzle input/output processor interface.

KIP Knowledge Industry Publications.

KIPO keyboard input printout.

KIPS 1. 10^3 (K) of instructions per second. Unit of computer processing speed. 2. knowledge information processing system. Systems using *AI*.

KIS keyboard input simulation.

KISS 1. keep it simple sir. 2. keyed indexed sequential search.

KIT 1. Kent Information Technology conference, part of *IT82*. 2. *Key issue tracking*. Current affairs database (US). 3. *KWIC* interactive tagger. Text editing system, University of Minnesota (US).

KKF *Magnetbanddienst Kunststoffe kautschuk fasern*. Database on plastics, *DKI*.

KL key length.

KLIC key letter in context. Indexing method.

KLU key and lamp unit, telecommunications.

KMON keyboard monitor, *DEC*.

KNAW Koninklijke Nederlandes Akademie van Wetenschappen (Royal Netherlands Academy of Arts and Sciences). Information broker.

KNO Koch, Neff, Oetlinger. Book wholesalers and developers of *BESSY* teleordering system (FRG).

KOBAS Konstanzer Bibliotheks-automatisierungs-system. Automated cataloguing system using *BAS* (FRG).

KOEBES Koelner Bibliotheksschliessungs-system. Automated library system, *HBZ*.

KOPS K (10^3) operations per second.

KORSTIC Korea Scientific and Technical Information Centre. *INSPEC* operator.

KP 1. key pulsing. 2. key punch.

KPC keyboard printer control.

KPH keystrokes per hour.

KPIC key phrase in context. Indexing method.

KPO key punch operator.

KR key register.

KRAS *Keyworded references to archaeological science*. Database, Department of Archaeology, University of Leicester (UK).

KRI King Research Inc. *IT* research and consultancy (US).

KRL knowledge representation language.

KRM Kurzweil reading machine. Uses *OCR* and voice synthesis.

KRN Knight Ridder Newspapers. Videotex producer, Viewdata Corp. (US).

KSAM 1. keyed sequential access method. 2. key field sequential access method, *HP*.

KSH key strokes per hour.

KSR keyboard send/receive.

KTBL *Kuratorium für Technik und Bauwesen in der Landwirtschaft*. Originator and databank on farm management (FRG).

KTDS key to disc software.

KTK Kommission des Technischen Kommunicationssystems (GDR).

KTM key transport module.

KTP keyboard typing perforator.

KTR keyboard typing reperforator.

KTS key telephone system. Manual switching unit.

KW kilo word.

KWAC key word and context. Indexing method.

KWADE key word as a dictionary entry. Indexing system, *IBM*.

KWIC key word in context. Indexing method.

KWIP key word in permutation. Indexing method.

KWIT key word in title. Indexing method.

KWOC key word out of context. Indexing method.

KWOT key word out of title. Indexing method.

KWUC key word and *UDC*. Indexing method.

KYBD keyboard.

L

L symbol for inductance.

LA 1. language. Searchable field, Dialog *ESA-IRS, NLM* and *SDC*. 2. The Library Association (UK). 3. line adaptor. 4. loan amount. Searchable field, Dialog *IRS*. 5. local address. 6. logical address.

LAAS Automation and Systems Analysis Laboratory, *CNRS*.

LABORDOC *Labour documentation*. Database on labour issues, *ILO*.

LAC 1. Library Advisory Council. Department of Education and Science (UK). Now *LISC*. 2. Library Assistants Certificate. City and Guilds Institute (UK).

LACAP Latin American cooperative acquisition project.

LACE Library Advisory Council for England. Now *LISC*.

LACES London Airport cargo *EDP* system.

LAD logical aptitude device.

LADB *Laboratory animal database*. Databank, *NLM*.

LADDER language access to distributed data with error recovery.

LADS local area data set.

LADSIRLAC Liverpool and district scientific industrial and research library advisory council. Library cooperative scheme (UK).

LAF long address form.

LAFIS local authority financial information system, *ICL*.

LAG 1. Librarians Automation Group (Australia). 2. load and go.

LAHCG look ahead carry generator, computing.

LAIG *LA* Industrial Group.

LAIS library acquisitions information system, Library Automation Research and Consulting Association (US).

LAIT Library Association Information Technology Group (UK).

LALS LaGuardia automated library system, LaGuardia Community College (US).

LAM loop adder and multiplier, computing.

LAMA local automatic message accounting.

LAMIS local authority management information system, Leeds City Council and *ICL* (UK).

LAMP library resources in literature, art, music and philosophy, Jackson State College (US).

LAMSAC Local Authorities Management Services and Computer Committee (UK).

LAN 1. lateral access network. Informal information system. 2. local area network. Network linking computers over limited area.

LANCET Library Association, National Council for Educational Technology. Cooperated to produce LANCET rules for cataloguing non-book material.

LAP 1. link access procedure. 2. list assembly program.

LAPL Library Association Publishing Limited (UK).

LAR 1. *Library Association record*. 2. limit address register, computing.

LARC Library Automation Research and Consulting Association (US).

LARP local and remote printing. Word processing.

LARPS local and remote printing station.

LA RSIS LA Reference, Special and Information Section (UK).

LAS local address space.

LASER 1. light amplification by the stimulated emission of radiation. 2. London and South Eastern Library Region. Library cooperative organization and *PRESTEL IP* for library and community information (UK).

115

LASIE library automated systems information exchange (Australia).

LASP local attached support processor.

LASS logistics analysis simulation system.

LASSOS Library Automation Systems and Services Options Study. *BL R&DD* advisory committee.

LAU 1. line access unit. 2. line adaptor unit, modems.

LB 1. *Library bookseller* (US). 2. line buffer, computing. 3. logical block.

LBA local bus adaptor.

LBC local bus controller.

LBEN low byte enable.

LBN logic bucket number, for record location.

LBR 1. laser beam recorder. 2. laser beam recording.

LBT low bit test.

LC 1. Library of Congress (US). 2. Library of Congress card number. Searchable field, *SDC*. 3. Library of Congress classification. Searchable field, *BLAISE*. 4. licensing country. Searchable field, Dialog *IRS*. 5. line circuit, telecommunications. 6. line concentrator. 7. line connector. 8. line control.

LCA 1. line control adaptor. 2. local communications adaptor. 3. lower case alphabet.

L carrier *SSBSC FDM* series, Bell System.

LCB 1. line control block, telecommunications. 2. logic control block.

LCBS London classification of business studies. Library classification scheme (UK).

LCC 1. late choice call. 2. lost calls cleared, telecommunications.

LCCC Library of Congress computer catalog.

LCCM late choice call meter, telecommunications.

LCCMARC Library of Congress current *MARC* file. Database on monographs from 1977.

LCCCN Library of Congress Catalog Card Number.

LCD liquid crystal display.

LCDS low cost development system, National Semiconductor (US).

LCF logical channel fill.

LCFS last come first served.

LCH lost calls held, telecommunications.

LCL 1. limited channel log-out. 2. linkage control language.

LCM 1. large core memory. 2. last calls meter, telecommunications. 3. line concentrator module. 4. line control module.

LCMARC Library of Congress *MARC* files from 1968.

LCMS library collection management system, developed by *UTLAS*.

LCN logical channel number, in *PSS*.

LCP 1. language conversion program. 2. link control procedure, data communications. 3. local control point.

LCPC Laboratoire Central des Ponts Chaussées. *IRRD* operator (France).

LCS 1. large capacity (core) storage, computing. 2. library computer system, University of Illinois (US). 3. loadable control storage, computing.

LCSH Library of Congress subject headings.

LCT logical channel termination.

LCU line control unit, data communications.

LCW line control word.

LD 1. logical design. 2. long distance. 3. low density.

LDA 1. load in accumulator, assembler code. 2. local data administrator. 3. locate drum address. 4. logical device address, computing.

LDB logical database.

LDC 1. *Linguistics documentation center*. Database, University of Ottawa (Canada). 2. local display controller. 3. low speed data channel.

LDCS long distance control system, Datapoint (US).

LDD local data distribution.

LDDS 1. limited distance data set. Modem. 2. low density data system.

LDLA limited distance line adaptor.

LDM 1. limited distance modem. 2. linear delta modulation. 3. local data manager.

LDP language data processing.

LDR low data rate.

LDRI low data rate input.

LDS 1. large disc storage. 2. local distribution service, cable TV network.

LDT 1. language-dependent translator. 2. logical design translator. 3. long distance transmission.

LDX long distance xerography. Fax system, Xerox Corporation.

LE 1. less than or equal to. Relational operator. 2. local exchange, telecommunications. 3. logic element.

LEAD learn execute and diagnose.

LEADER Lehigh automatic device for efficient retrieval, Lehigh University (US).

LEADERMART *LEADER* mechanical analysis and retrieval of text. Information retrieval system.

LEAF *LISP* extended algebraic facility.

LEASAT leased satellite (US).

LEC London education classification. Library classification system.

LED light emitting diode. Display component.

LEDOC *Lehrstuhlinformationssystem.* Database on computers and law, University of Regensburg (FRG).

LEEP library education experimental project, University of Syracuse (US).

LEF line expansion function.

LEG Library Education Group of the Library Association. Now *TEG*.

LEGOL legally oriented language. Programming language project (UK).

LEM 1. logical end of medium. 2. logical enhanced memory.

LEMME Legislacao referente ao Ministerio das Minas e Energia. Indexing service on legal texts in mining and energy (Brazil).

LEO 1. Lyons electronic office. Early computer, J. Lyons & Co (UK). 2. computer series, *ICL*.

LETIS Leicestershire technical information service (UK).

LEX line exchange(r), telecommunications.

LEXINFORM *legal database, DATEV.*

LEXIS 1. database on law, not an acronym (US). 2. lexicography information service. Machine translation system (FRG).

LEXPAT Patent search and retrieval series, Mead Data Central (US).

LF 1. linear file. Computer file, *NASA* and *ESRO*. 2. line feed. Control character. 3. line finder, telecommunications. 4. logical file. 5. logic function. 6. low frequency (approx 10^5 Hz).

LFC local forms control, computing.

LFD local frequency distribution.

LFJ local feed junctor, telecommunications.

LFM local file manager.

LFN logical file name.

LFO low frequency oscillator.

LFPS low frequency phase shifter, telecommunications.

LFS local format storage.

LFSR linear feedback shift register.

LFU 1. least frequency unit. 2. least frequently used, computing.

LGN logical group number.

LHF list handling facility.

LI 1. Learned Information. Database originator and marketer, publisher of *IT* books and journals. 2. special list indicator. Searchable field, *NLM*.

LIA loop interface address.

LIBCEPT LIBRIS intercept. Processing system for *LIBRIS* (Sweden).

LIBCON/E *Library of Congress/English*. Database on English language monographs.

LIBRI *Literary information bases for research and instruction*. Greek and Latin full text database, American Philological Association.

LIBRICON computer and consulting service (libraries) Auto Graphics Inc.

LIBRIS 1. Leuvens integraal bibliotheck systeem. Library circulation system (Belgium). 2. *Library information system*. Database of books in Scandinavian academic libraries, *FBR* and *SAFAD*.

LICOS logical input/output control system, Olivetti (Italy).

LID 1. leadless inverted device. Semiconductor. 2. line isolation device, telecommunications. 3. Literaturdienst Medizin. Online information broker (Austria).

LIE Lessio Intellectuale Europeo. Research Institute, Consiglio Nationale delle Richerche (Italy).

LIFO last in first out.

LIFT logically integrated *FORTRAN* translator.

LIH line interface handler.

LILO last in last out.

LIM 1. language interpretation module. 2. line interface module.

LIMA logic in memory array.

LINC laboratory instrument computer.

LINCS language information network and clearing house system, Center for Applied Linguistics (US).

LINK Lambeth information network. Information service for industry and commerce (UK).

LINOSCO libraries in North Staffordshire and South Cheshire in cooperation. Library cooperation scheme (UK).

LINUS logical inquiry and update system.

LIOCS logical input/output control system.

LIOP logical input/output processor.

LIPL linear information programming language.

LIPS logical inferences per second.

LIRES-MS literature retrieval system – multiple searching complete text. Institute of Paper Chemistry (UK).

LIRG Library and Information Research Group (UK).

LIRN *Library and information research news*. Newsletter, *LIRG*.

LIS 1. legislative information system, National Conference of State Legislatures (US). 2. library and information science. 3. List and Index Society (UK). 4. Lockheed Information Systems. Host (US).

LISA 1. *Library and information science abstracts*. Journal and database, *LAPL*. 2. linked indexed sequential access. 3. locally integrated software architecture, Apple microcomputer.

LISARD 1. library and information service automated retrieval of data. 2. library information search and retrieval data system, US Navy.

LISC Library and Information Service Council. Advisory body (UK). Formerly *LAC*.

LISD Library and Information Services Division, National Oceanic and Atmospheric Administration (US).

LISIC library and information service to industry and commerce. Now incorporated in *SEAL*.

LISP list processing. Programming language.

LISR line information storage and retrieval. Information system, *NASA*.

LIST Library Information Service for Teesside. Library cooperative (UK).

LISTAR Lincoln information storage and associative retrieval system. Lincoln Laboratory, Massachusetts Institute of Technology (US).

LIT light interchange (interface) technology. Interface between optical and electronic components.

LITA Library and Information Technology Association of the *ALA*. Formerly *ISAD*.

LITIR *Literature information and retrieval.* Database on Victorian studies literature, University of Alberta (Canada).

LIT-KRAN *Literaturdokumentation Krankenhausen.* Medical database. *DKI* and Technische Universität Berlin (FRG).

LIU line interface unit, data communications.

LJE local job entry.

LKM low key maintenance.

LL local line.

LLA leased line adaptor.

LLBA *Language and language behavior abstracts.* Database, Sociological Abstracts Inc. (US).

LLC logic link control, network interfacing.

LLE long line equipment, telecommunications.

LLG logical line group.

LLI low level interface.

LLL low level language.

LLLLL laboratories low level linked list.

LLM low level multiplexer.

LLN 1. line link network. 2. local line network, telecommunications.

LLR Lloyds Register of Shipping. Originates and operates databanks on shipping (UK).

LLS local library system, *OCLC*.

LM 1. limitation. Searchable field, Dialog *IRS*. 2. link manager. 3. load module. 4. local memory. 5. logical module. 6. loop multiplexer.

LMBI local memory bus interface, computing.

LMI local memory image.

LML logical memory level.

LMRU Library Management Research Unit, Cambridge University (UK). Now *CLAIM*.

LMS 1. level measuring set, for test signals. 2. library management system, University of Dortmund and *IBM*. 3. list management system.

LMT logical mapping table.

LMU line monitor unit.

LNA low noise amplifier.

LNB local name base, computing.

LNC low noise converter.

LND local number dialling, telecommunications.

LNE local network emulator.

LNR low noise receiver.

LO 1. local oscillator. 2. location code. Searchable field, Dialog *IRS*. 3. low, of voltage. 4. lowest grant. Searchable field, Dialog *IRS*.

LOA log-out analysis, Fujitsu (Japan).

LOC location. Searchable field, Pergamon Infoline.

LOCAL load on call. Computing.

LOCAS local cataloguing service. *BLAISE* service.

LOCATE Library of Congress automation techniques exchange.

LOCS logic and control simulation.

LOEX Library Orientation Instruction Exchange. National clearing house for library user instruction (US).

LOFAR low frequency analysis recording.

LOGEL logic generating language.

LOGFED log file editor.

LOGIC programming language, Datasaab (Sweden).

LOGIPAC logical processor and computer.

LOGOL language oriente gestion des ordinateurs LogAbax. Computer language, LogAbax (US).

LOLA 1. library online acquisitions, Washington State University (US). 2. London online local authorities (UK).

LOLC library developed online catalogue.

LOLITA library online information and text access, Oregon State University (US).

LOMAC low level macroprocessor language, Business Computers (US).

LOP line-oriented protocol.

LOS 1. line of sight. 2. loss of signal.

LOSR limit of stack register.

LOT light-operated typewriter.

LOTIS logical timing sequencing.

LOW low core threshold.

LP 1. light pen. 2. linear programming. 3. linear programming language, Intertechnique (France). 4. line printer. 5. logic probe. 6. log periodic, antenna.

LPA linear power amplifier.

LPC linear predictive coding. Digital coding technique.

LPCM linear phase code modulation.

LPD language processing and debugging.

LPF low pass filter.

LPI lines per inch.

LPID logical page identifier.

LPL 1. linear programming language. 2. list processing language. 3. local processor link.

LPM lines per minute.

LPN logical page number.

LPS 1. laser printing system, Wang. 2. linear programming system. 3. lines per second.

LPTTL low power transistor to transistor logic.

LPU 1. language processor unit. 2. line processing unit.

LPVT large print video terminal.

LR 1. limit register. 2. logical record. 3. low reduction (less than 15x), microfilm.

LRA logical record address.

LRC longitudinal redundancy check.

LRCC Library Resources Coordinating Committee of the University of London (UK).

LRI *Legal resource index*. Database, Information Access Co. (US).

LRL 1. linking relocating loader. 2. logical record length. 3. logical record location.

LRR loop regenerative repeater.

LRU least recently used.

LS 1. local store. 2. low speed.

LSA line sharing adaptor.

LSAA Library Services Authority Act (US).

LSA mode limited space charge accumulation mode, telecommunications.

L-SAT communications satellite project, *ESA*.

LSB 1. least significant bit, data compaction. 2. lower side band, data transmission.

LSC 1. local switching centre, telecommunications. 2. low speed concentrator.

LSD 1. language for systems development. 2. least significant digit, data compaction. 3. line signal detector. 4. low speed data.

LSDR local store data register.

LSE London School of Economics. Database originator and operator (UK).

LSFR local storage function register.

LSG Language Structure Group, *CODASYL*.

LSI 1. large scale integration, electronics. 2. Lear Siegler Inc. (US).

LSIB *London stage information bank*. Database on 18th century theatre.

LSID local session identification.

LSL 1. ladder static logic. 2. link and selector language.

LSLA low speed line adaptor, data communications.

LSM line select module.

LSMA low speed mutliplexer arrangement.

LSP 1. library software package, *LOCAS*. 2. linked systems project, *WLN, RLG* and LC. 3. local store pointer.

LSQA local system queue area, computing.

LSR 1. local shared resources. 2. local storage register. 3. low speed reader.

LSS 1. language for symbolic simulation. 2. loop switching system, telecommunications.

LST loud speaking telephone.

LSTB low power Schottky *TTL* bipolar. *LSI* technology.

LS-TTL-LSI low power Schottky transistor to transistor logic *LSI*.

LSU 1. library storage unit. 2. line selection unit, telecommunications. 3. line sharing unit. 4. load storage unit. 5. local storage unit. 6. lone signalling unit, telecommunications.

LT 1. language translation. 2. less than. 3. line terminator. 4. logic theory.

LTA logical transient area.

LTB last trunk busy, telecommunications.

LTC 1. line time clock. 2. local terminal controller.

LTD line transfer device.

LTE local telephone exchange.

LTH logical track header.

LTM long-term memory.

LTPL long-term procedural language.

LTRS letter shift.

LTS line transient suppression.

LTU line termination unit.

LU logical unit.

LUB logical unit block.

LUCK logical unit and checker, Amdahl (US).

LUE link utilization efficiency.

LUF lowest usable frequency, in *HF* propagation.

LUG *LOCAS* Users Group (UK).

LUIS library user information system. Online cataloguing system, North Western University (US).

LULU logical unit to logical unit.

LUN logical unit number.

LUTFCSUSTC Librarians United to Fight Costly Silly Unnecessary Serial Title Changes. Pressure group (US).

LUTIS Luton Information Service. Library cooperative (UK).

LVA local virtual address.

LVLSH level shifter, computing.

LVR longitudinal video recording.

LWA last word address.

LXMAR load external memory address register.

M

m milli. One thousandth (10^{-3}).

M mega. One million (10^6).

MA 1. memory address. 2. message assembler. 3. microfilm address.

MAACS multi-address asynchronous communication system.

MAB 1. macro-address bus. 2. Maschinelles Austauschformat für Bibliotheken. Collaborative library automation project (FRG).

MAB1 MAB version 1. Machine-readable data format.

MAC 1. machine-aided cognition. 2. measurement and analysis centre. 3. media access control, network interfacing. 4. memory access controller. 5. message authentication code. 6. multi-access computing. 7. multiple access computer. 8. multiplexed analog components, for satellite broadcasting.

MACCS molecular access system, for chemical structure, Shell Research Ltd.

MACDAC machine communication with digital automatic computer.

MACDACsys *MACDAC* system.

MACROCAL enhanced macro version of common assembler language, Interdata (US).

MACROL macro-based display oriented language, Raytheon (US).

MACS 1. metering and accounting system. Automated telecommunications accounting, Plessey (UK). 2. monitoring and control station.

MAD 1. machine *ANSI* data. 2. Michigan algorithmic decoder. Computer programming language. 3. multiple access drive.

MADA multiple access discrete address.

MADAM multi-purpose automatic data analysis machine.

MADAR malfunction analysis detection and recording.

MADDIDA magnetic drum digital differential analyser.

MADE micro-alloy diffused electrode.

MADM Manchester automatic digital machine. First operational stored program computer, Manchester University (UK).

MADR microprogram address register.

MADRE magnetic drum receiving equipment.

MADT micro-alloy diffused base transistor.

MAE memory address register.

MAG 1. macro-generator, *SEMIS*. 2. magnetic.

MAGB Microfilm Association of Great Britain.

MAGIC 1. machine-aided graphics for illustration and composition. 2. matrix algebra general interpretive coding.

MAGTC magnetic tape controller.

MAI 1. machine-aided index. 2. multiple access interface.

MAL 1. macro-assembly language. 2. memory access logic. 3. meta-assembly language.

MALCAP Maryland Academic Library Center for Automated Processing, University of Maryland (US).

MALMARC Malaysian *MARC*.

MAM 1. memory access multiplexer. 2. memory allocation manager. 3. message access method. 4. multi-application monitor. 5. multiple access to memory.

MANDATE multi-line automatic network diagnostic and transmission equipment.

MANIAC mathematical analyser numerical integrator and computer.

MANMAX machine-made and machine-aided index.

123

MANTIS Manchester technical information and commercial service (UK).

MAP 1. macro-assembly program. 2. memory allocation and protection. 3. memory allocation processor. 4. message acceptance pulse. 5. Microprocessor Application Project, in manufacturing industry, *DoI*. 6. microprogrammed array processor. 7. model and program. 8. modular analysis processor. 9. multiple allocation procedure.

MAPCON Microprocessor Applications Consultancy. Scheme sponsored by *DoI*.

MAPPER maintaining, preparing and producing executive reports. Univac software for inexperienced users.

MAPS management accounting and payroll system, Plantime (UK).

MAPTEL Maplin Telecommunications. Microcomputer network, Maplin Company (UK).

MAR 1. macro-address register. 2. memory address register. 3. microprogram address register. 4. miscellaneous apparatus rack.

MARC machine-readable catalog. Catalog format developed by *LC*.

MARC IS *MARC* Israel. *MARC* for English books (Israel).

MARCIVE *MARC* five. *MARC*-based system at five academic libraries, San Antonio (US).

MARC(LC) *MARC* Library of Congress. Database on *LC* holdings.

MARCS *MELCOM* all round adaptive consolidated software, Fujitsu (Japan).

MARC(S) *MARC* serials.

MARC(UK) *MARC* United Kingdom. Database on *BL* holdings.

MARECS marine communications satellites. Developed by *ESA*.

MARGIE memory analysis response generation and interference in English.

MARISAT maritime satellite system. Operated by *COMSAT*.

MARLIS multi-agent relevance linkage information system.

MARNA *Marine navigation*. Database, Samson Data Systemen (Netherlands).

MARS 1. Machine-Assisted Reference Section, Reference and Adult Services Division, *ALA*. 2. machine retrieval system. 3. multiple access retrieval system.

MARSL machine-readable shelf list, Carleton University (Canada).

MARTOS multi-access real time operating system, *AEG* Telefunken (FRG).

MARVEL managing resources for university libraries. Integrated library system, Georgia University (US).

MAS 1. macro-assembler. 2. modular application system.

MASCOT modular approach to software construction operation and test.

MASER microwave amplification by the stimulated emission of radiation.

MASM 1. macro-assembler. 2. meta-assembler, Sperry Univac (US).

MASS 1. *MARC-based automated serials system*. Database, *BLCMP*. 2. multiple access sequential selection. 3. multiple access switching system.

MASSBUS memory bus, *DEC*.

MASTER multiple access shared time executive routine, *CDC*.

MASTIR microfilmed abstracts system for technical information retrieval, Illinois Institute of Technology (US).

MAT 1. machine-aided translation. 2. memory address test. 3. memory address translator. 4. micro alloy transistor.

MATE 1. memory-assisted terminal equipment, for disabled users, Essex University. 2. modular automatic test equipment.

MATR management access to records.

MATV master antenna television.

MAU 1. medium access unit, network interfacing. 2. memory access unit. 3. multiple access unit.

MAVICA magnetic video card. Sony Company (Japan).

MAX 1. maximum. 2. mobile automatic exchange, telecommunications. 3. modular applications executive, Modular Computer Systems (US).

124

MAXCOM modular applications executive for communications, Modular Computer Systems (US).

MAXNET modular applications executive network. Operating system, Modular Computer Systems (US).

MB 1. megabit. 2. megabyte. Units of data storage capacity. 3. memory buffer. 4. memory bus.

MBANK *Kentucky monthly state databank*. Databank, *KEIS*.

MBC 1. memory bus controller. 2. multiple basic channel.

MBCD modified binary-coded decimal.

MBD manual board, telecommunications.

MBI memory bank interface.

MBIO microprogrammable block input/output.

MBM magnetic bubble memory.

MBOS multi-user business operating system.

MBPS megabits per second.

MBR 1. memory base register. 2. memory buffer register.

MBS 1. megabits per second, transmission rate. 2. multiple batch station, computing.

MBU 1. magnetic bubble unit. 2. memory buffer unit, computing.

MC 1. magnetic card. 2. main channel. 3. manual code. Searchable field, Pergamon Infoline. 4. marginal checking. 5. master control. 6. measure code. Searchable field, Dialog *IRS*. 7. memory control. 8. *MESH* class number. Searchable field, *NLM*.

MCA 1. Microfilm Corporation of America. 2. multiprocessor communications adaptor.

MCAI microcomputer-assisted instruction.

MCALS Minnesota computer-aided library system, University of Minnesota (US).

MCAR machine check analysis and recording.

MCB microcomputer board.

MCBF mean cycles between failures.

MCC 1. maintenance control centre. 2. Microelectronics and Computer Technology Corporation (US). 3. mini-channel communications control. 4. miscellaneous common carrier. 5. multichannel communications controller. 6. multiple chip carrier.

MCCU 1. multiple channel control unit. 2. multiple communications control unit.

MCDBSU master control and data buffer storage unit.

MCEL machine check extended log-out.

MCF museum communication format, Information Retrieval Group of the Museums Association (UK).

MCG man computer graphics.

MCH 1. machine check handle, Fujitsu (Japan). 2. machine check-handler.

M chs thousands (10^3) of characters.

MCI 1. machine check interrupt, computing. 2. Microwave Communications Inc. (US).

MCIC machine check interruption code.

MCL 1. memory control and logging. 2. microprogram control logic.

MCLA 1. micro-coded communications line adaptor. 2. micro-coded communications link adaptor.

MCM memory control module.

MCN 1. Micro Networks Corporation (US). 2. Museum Computer Network (US).

MCOS microprogrammable computer operating system.

MCP 1. master control program, Burroughs (US). 2. memory centred processor. 3. message control program. 4. multi-channel communications program.

MCPG media conversion program generator.

MCPU multiple central processing unit.

MCR 1. magnetic card reader. 2. master control register. 3. memory control register.

MCRN Moscow City Relay Network.

MCRR machine check recording and recovery.

125

MCRS micro-graphics catalog retrieval system. Information retrieval system, *LC*.

MCS 1. marine communication subsystem, *INTELSAT*. 2. master control system. 3. message control system, Burroughs (US). 4. microcomputer system. 5. Mini-Computer Systems Inc. (US). 6. multi-console system, *SEMIS*. 7. multi-channel communications software. 8. multiple console support, Fujitsu (Japan).

MC/ST magnetic card selectric typewriter, *IBM*.

MCT mobile communication terminal.

MCU 1. maintenance control unit, computing. 2. master control unit. 3. memory control unit. 4. microcomputer control unit. 5. micro-control unit, Pertec (US). 6. microprocessor control unit. 7. microprogram control unit, computing. 8. multiplexer control unit. 9. multiprocessor communications unit. 10. multi-system communications unit.

MCVD modified chemical vapour deposition. Technique in fibre optics manufacture.

MCVF multi-channel voice frequency.

MCW modulated continuous wave, signal transmission.

MD 1. message data. 2. minidisk.

MDA 1. multi-dimensional access. 2. multi-dimensional array.

MDAC multiplying digital-to-analog converter.

MDC 1. Machinability Data Center. Originator and its databank (US). 2. Mead Data Control. Produces *LEXIS* and *NEXIS* (US). 3. memory disc controller. 4. Microprocessor Development Center, American Microsystems Inc. (US). 5. multiple device controller.

MDCU magnetic disc control unit.

MDD magnetic disc drive.

MDE magnetic decision element.

MDES multiple data entry system.

MDF main distribution frame. Telecommunications interface.

MDF/1 *Metals data file/1*. Database, *ASM*.

MDI medium-dependent interface, network interfacing.

MDL 1. macro-description language. 2. maintenance and diagnostic logic display, Burroughs (US).

MDLC multiple data link controller.

MDM multiplexer/demultiplexer.

MDMS multiple database management system.

MDO *MARC* Development Office, *LC*.

MDOS Motorola disc operating system.

MDP 1. main data path. 2. maintenance diagnostic processor.

MDPHI media development project for the hearing impaired. Videodisc project (US).

MDPS multiple workstation direct processing system, *NCR*.

MDR 1. magnetic disc recorder. 2. Market Data Retrieval. Originator, operator and databank (US). 3. memory data register. 4. multi-disc reader, for floppy discs.

MDS 1. microprocessor development system, Motorola (US). 2. modular data system. 3. modular disc storage. 4. Mohawk Data Systems Corporation (US). 5. multipoint distribution service. Microwave TV service. 6. multiprocessor distributed system, Raytheon (US).

MDSS microprocessor development support system.

MDT 1. mean down time. 2. multi-dimensional tasking.

MDU maintenance diagnostic unit.

ME 1. main entry (author). Searchable field, *SDC*. 2. memory element, computing.

MEB modem evaluation board.

MECHEN *Mechanical engineering*. Database, *RITL*.

MEDAC medical electronic data acquisition and control.

MEDI *Marine environmental data information retrieval system*. Databank originated and operated by *IOC*.

MEDIA/M *Media/medicine.* Databank on readership of French medical media, *CESSIM.*

MEDIA/P *Media/Publicitaire.* Databank on readership data of French publicity media, *CESP.*

MEDICO model experiment in drug indexing by computer, Rutgers University (US).

MEDICS medical information and communications system.

MEDL Marconi Electronic Devices Ltd. (UK).

MEDLARS *Medical literature analysis and retrieval system.* Database, *NLM.*

MEDLINE medical information online, *MEDLARS* information retrieval system.

MEDOC *Medical documents.* Database, University of Utah (US).

MEDTRAIN *Medical literature training file.* Database for search training, *NLM.*

MEF *MidEast file.* Database, Learned Information (UK).

MEI *Main economic indicators.* Databank, OECD.

MEIS *Military entomology information service.* Database, US Department of Defense.

MELCOM computer series, Mitsubishi (Japan).

MELCU multiple external line control unit.

MELOP million floating point operations per second. Measure of computing power, also *MFLOPS.*

MELVYL name for Melvil Dewey. Public access online catalog, University of California (US).

MEME multiple entry multiple exit.

MEP 1. micro-electronics education programme. Government programme for schools (UK). 2. microfile enlarger printer.

MERLIN machine-readable library information. *BL* project cancelled 1979.

MESH medical subject headings. *MEDLARS* thesaurus.

META 1. assembler, *CDC.* 2. computer series, Digital Scientific (US).

METADEX *Metals abstracts index.* Database, American Society for Metals, The Metals Society (UK).

METAPLAN methods of extracting text automatically programming language. Text retrieval language.

MEU memory expansion unit.

MEWT matrix electrostatic writing technique.

MF 1. master file. 2. microfarad. 3. medium frequency (approx 10^6 Hz). 4. microfiche. 5. microfilm. 6. microform. 7. molecular formula. Searchable field, Dialog *IRS.* 8. multi-frequency.

MFCA multi-function communications adaptor.

MFCM multi-function card machine.

MFCU multi-function card unit.

MFD master file directory.

MFDSUL multi-function data set utility language.

MF4 multi-frequency signalling service, *BT.*

MFG 1. message flow graph. 2. multi-function generator.

MFLOPS million floating point operations per second. Measure of computing power, also *MELOP.*

MFLP multi-file linear programming.

MFM 1. modified frequency modulation. 2. multi-stage frequency multiplexer.

MF/1 measurement frequency/1, *IBM.*

MFP 1. multi-form printer. 2. multiprogramming fixed task, *IBM.*

MFPC multi-function protocol converter.

MFR 1. manufacturer. 2. multi-frequency receiver, telecommunications.

MFS multi-frequency signalling, telecommunications.

MFSK multiple frequency shift keying.

MFT multiprogramming, fixed tasks.

MG motor generator.

MGA *Meteorological and geoastrophysical abstracts*. Database, American Meteorological Society (US).

MGL matrix generator language.

MGP multiple goal programming.

MH 1. manual hold, telecommunications. 2. *MESH* heading. Searchable field, *NLM*. 3. message handler, computing.

M-H McGraw-Hill. Electronic publisher (US).

MHC modified Huffman coding. Data compaction algorithm.

MHD 1. movable head disc. 2. moving head disc. 3. multiple head disc.

MHP message handling processor.

MHS multiple host support.

MHSDC multiple high speed data channel.

MHz megahertz. One million *Hz*.

MI 1. machine independent. 2. *Magazine index*. Database, *IAC*. 3. manual input. 4. memory interface. 5. *Monitoring information*. Databank, *ITU*.

MIA multiplex interface adaptor.

MIACF meander inverted autocorrelated function.

MIAS *Marine information and advisory service*. Databank on tides and waves, Institute of Oceanographic Sciences (UK).

MIB micro-instruction bus.

MIC 1. Maruman Integrated Circuits. Microprocessor manufacturer (US). 2. Medical Information Centre. Host (Sweden). 3. missing interruption character.

MICA 1. macro-instruction compiler assembler. 2. major incidents computer application. Police system (UK).

MIC-KIBIC Medinska Informationcentralen – Karolinska Institutetsbibliotek och Informationscentralen. Host, database originator and information broker (Sweden).

MICOS Mini Computer Systems. Computer manufacturer (US).

MICR magnetic ink character recognition. Computer input.

MICRO 1. microcomputer. 2. multiple indexing and console retrieval operations. Information retrieval system.

MICROCAT micro-catalogue. Very short entry catalogue, University of Southampton (UK).

MICROM micro-instruction read-only memory.

MICRONET microcomputer network. *PRESTEL* database (UK).

MICROPSI microcomputer printed subject indexes, *CLW*.

MICS management information and control system.

MID message input description.

MIDAS 1. medical information dissemination using *ASSASSIN, ICI*. 2. memory implemented data acquisition systems. 3. micro-diagnostics for analysis and repair. 4. microprogrammable integrated data acquisition system. 5. microprogramming design-aided system. 6. modified integrated digital to analog simulator. 7. modular international dealing and accounting system, *BIS* Software Ltd. (UK). 8. multi-mode international data acquisition service. Information search network (Australia). 9. multiple index data access system.

MIDDLE microprogram design and description language.

MIDEF micro-procedure definition.

MIDIST Le Mission Interministérielle de l'Information Scientifique et Technique. National authority for scientific and technical information (France).

MIDLNET Midwest Library Network. Library cooperative (US).

MIDMS machine-independent data management system.

MIDS multi-mode information distribution system.

MIFR *Monitored international frequency register*. Databank, *ITU*.

MIH 1. missing interruption handler, Fujitsu (Japan). 2. multiplex interface handler.

MIKAS Mikrofiche-katalog-system. Cataloguing system, *ETH*.

MIL 1. micro-implementation language, Burroughs (US). 2. module interconnection language.

MILO Maryland interlibrary loan. *ILL* system (US).

MIL-STD military standard. Applied to many US electronic devices.

MIM 1. maintenance interface machine, *CIIHB*. 2. *MODEM* interface module.

MIMC Microforms International Marketing Company, Pergamon.

MIMD 1. multiple instruction/multiple data. 2. multiple instruction stream multiple data stream. Parallel processor configuration.

MIMO man in/machine out.

MIMOLA machine-independent microprogramming language.

MIN 1. minimum. 2. minute.

MIND modular interactive network designer.

MINERVE méchanisation de l'information dans l'exploration et la recherche verrières. Institut du Verre (France).

MINICS minimal input cataloguing system. Document cataloguing system.

MINICS/PDS *MINICS* periodicals data system.

MINIPAC minicomputer control program and highway data controller, *ICL*.

MINITEX Minnesota Interlibrary Telecommunications Exchange. Library cooperative (US).

MINOS modular input/output system.

MIO multiple input/output.

MIOP 1. master input/output processor. 2. multiplexer input/output processor.

MIOS modular input/output system.

MIP 1. machine construction processor. 2. manual input processing. 3. matrix inversion program. 4. mixed integer programming.

MIPE modular information processing equipment.

MIPS millions of instructions per second. Measure of computing power.

MIR 1. memory information register. 2. memory input register. 3. micro-instruction register. 4. music information retrieval.

MIRPS multiple information retrieval by parallel selection.

MIRS *Medical information retrieval service*. Database for use with videodisc for graphics, W. B. Saunders (US).

MIS 1. management information system. 2. *Marine information system*. Databank, Marine Management Systems Inc. (US). 3. *Mathematik informationssystem*. Database, *ZFM*. 4. metal insulator semiconductor.

MISAM multiple index sequential access method.

MISD multiple instruction stream single data stream. Parallel processor configuration.

MISER minimum size executive routines.

MISLIC Mid- and South Staffordshire Libraries in Cooperation. Library cooperative (UK).

MISMDS multiple instruction streams multiple data streams.

MISP medical information systems program.

MISSIL management information system symbolic interpretive language.

MIST music information system for theorists. Program package, Indiana University (US).

MIT 1. Massachusetts Institute of Technology (US). 2. master instruction tape. 3. Mitsubishi Electric Corporation (Japan). 4. modular industrial terminal. 5. modular intelligent terminal.

MITA Microcomputer Industry Trade Association.

MITI Ministry of International Trade and Industry (Japan).

MITOL machine-independent telemetry oriented language.

MITS 1. management information and text system. 2. Micro Instrumentation and Telemetry Systems. Home computer kit manufacturer (US).

MIU model interface unit.

MIW micro-instruction word.

MKH multiple key hashing.

MKR marker, in switching system.

MKS metre kilogram second. System of units.

ML 1. machine language. 2. manipulation language. 3. microprogramming language.

MLA 1. matching logic and adder. 2. Modern Language Association of America. Originates *MLA bibliography* database. 3. multiple line adaptor.

MLC 1. Michigan Library Consortium. Library cooperative (US). 2. microprogram location counter. 3. multi-line control.

MLCP multi-line communications processor.

MLD machine language debugger, National *CSS*.

MLE Merrill Lynch Economics. Originator and databank (US).

MLI 1. machine language instruction. 2. message level interface.

MLIA multiplex loop interface adaptor.

MLIM matrix log-in memory.

MLIP message level interface port.

MLP machine language program.

MLS 1. machine literature searching. 2. multi-language system.

MLTA multiple line terminal adaptor.

MLU 1. memory logic unit. 2. multiple logic unit.

MM 1. mass memory. 2. memory module.

MMA 1. *Management and marketing abstracts*. Database, *PIRA*. 2. multiple module access.

MMAR main memory address register.

MMC multipart memory controller.

MMCP micro-master control processor.

MMDS Martin Marietta Data Systems (US).

MMF magnetomotive force.

MMFA modified modified frequency modulation.

MMFM modified modified frequency modulation. Also *M2FM*.

MMI 1. main memory interface. 2. man-machine interaction. 3. man-machine interface. 4. Monolithic Memories Incorporated. Microprocessor manufacturer (US). 5. multi-message interface.

MMIU multi-part memory interface unit.

MML man-machine language.

MMOS message multiplexer operating system.

MMR 1. magnetic memory record. 2. main memory register.

MMS 1. man-machine system. 2. mass memory store. 3. memory management system. 4. microfiche management system. 5. Micro Memory Systems. Computer manufacturer (UK). 6. multi-part memory system, Perkin-Elmer (US).

MMU 1. main memory unit. 2. memory management unit. 3. memory mapping unit.

MN 1. manufacturer's name. 2. measurement name. 3. meeting number. Searchable fields, Dialog *IRS*. 4. meetings name. Searchable field, *BLAISE*.

MNA multi-share network architecture, Mitsubishi (Japan).

MNCS multi-point network control system.

MNDX mobile non-director exchange, telecommunications.

MNF multi-system networking facility.

MNMKT *Money market*. Databank, *CSC*.

MNOS metal-nitride-oxide-silicon. Type of *EAROM*.

MNSC main network switching centre.

MO memory output.

MOB basic operating monitor, *SEMIS*.

MOBIDAC mobile data acquisition.

MOBL macro-oriented business language. Programming language.

MOBOL Mohawk business-oriented language. Mohawk Data Systems (US).

MOD message output description.

MODACS modular data acquisition and control system, *MODCOMP*.

MODCOMP Modular Computer Systems Inc. (US).

MODEM modulator-demodulator. Telecommunications.

MODI modular optical digital interface.

MOJ metering over junction, network administration.

MOL machine-oriented language.

MOLARS Meteorological Office library accessions and retrieval system (UK).

MOLDS management online data system, University of Syracuse (US).

MONOCLE projet de mise en ordinateur d'une notice catalographique de livre. Processing format using *MARC* and French cataloguing practice.

MOP 1. multiple online processing. 2. multiple online programming.

MOPS million operations per second. Units of processing power.

MOR memory output register.

MOS 1. management operating system. 2. master operating system. 3. mathematical offprint service, *AMS*. 4. memory-oriented system. 5. metal oxide semiconductor. 6. microprogram operating system. 7. minimum operating system, Sperry Univac (US). 8. modular operating system. 9. multiprogramming operating system.

MOSAIC macro-operation symbolic assembler and information compiler.

MOSFET metal oxide semiconductor field effect transformer.

MOTNE Meteorological Operational Telecommunications Network Europe.

MOUG Maryland Online User Group (US).

MOUTH modular output unit for talking to humans.

M out of N error detection code, telecommunications.

MP 1. macroprocessor. 2. metra potential. 3. microprogram. 4. multiprocessing. 5. multiprocessor.

MPA 1. multiple peripheral adaptor. 2. multi-point asynchronous.

MPAR microprogram address register.

MPC 1. microprogram control. 2. modular peripheral interface converter.

MPCC multi-protocol communications controller.

MPCI multi-port programmable communications interface.

MPCM microprogram control memory.

MPDC Mechanical Properties Data Center. Originator and database (US).

MP-DV multiply-divide.

MPE 1. memory parity error. 2. multiprogramming executive, *HP*.

MPF Micro Professor. Microcomputer claimed to be compatible with Apple II.

MPGS microprogramming generating system.

MPI 1. Magnetic Peripherals Inc. Manufacturer (US). 2. microprocessor interface.

MPL 1. message processing language, Burroughs (US). 2. microprogramming language. 3. multi-schedule private line.

MPLA mask programmable logic array.

MPM 1. microprogram memory. 2. multiprogramming monitor.

MP/M multiprogramming control program for microprocessors. Operating system, Digital Research.

MPMC microprogram memory control.

MPO memory protect override.

MPP 1. massively parallel processor. Used in image processing. 2. memory parity and protect.

MPPL 1. multi-purpose processing language. 2. multi-purpose programming language.

MPR message processing region.

MPROM mask programmable *ROM*.

MPS 1. microprocessor system. 2. multiprogramming system. 3. multiprogramming periodic tasking system, Raytheon (US).

MPSX mathematic programming system extended, *IBM*.

MPT 1. memory processing time. 2. Ministry of Post and Telecommunications (Japan).

MPU 1. memory protection unit. 2. microprocessor unit. *CPU* of microcomputer.

MPX 1. mapped programming executive, Systems Engineering Laboratories (US). 2. multiplexer. 3. multiprogramming executive.

MPX/R relay multiplexer.

MPX/S solid state multiplexer.

MQ multiplier/quotient.

MQE message queue element.

MQEM *Michigan quarterly economic model*. Databank, University of Michigan (US).

MQS multiprogrammed queued tasking system, Raytheon (US).

MR 1. mask register. 2. medium reduction, of microfilm (16x to 30x). 3. modular redundancy. 4. multiple request. 5. multiplier register.

MRA Machine Readable Archives Division, Public Archives of Canada.

MRAD mass random access disc.

MRB *Microcircuit reliability bibliography*. Database, Rome Air Development Center (US).

MRC 1. machine-readable code. 2. Medical Research Council. Database originator and operator (UK). 3. memory request controller.

MRCS multiple report creation system.

MRDF machine-readable data files.

MRDOS mapped real time disc operating system.

MRF message refusal signal, telecommunications.

MRI memory reference instruction.

MRIS Maritime Research Information Service. Originator and databank, division of *NAS*.

MRJE multiple remote job entry.

MRL machine representational language.

MROM macro *ROM*.

MRR multiple response resolver.

MRS management reporting system.

MRT mean repair time.

MRWC multiple read write compute.

MS 1. main storage. 2. mark sense. 3. mass storage. 4. millisecond.

MSA 1. Management Science America, owns Peachtree Software. 2. mass storage adaptor. 3. multi-subsystem adaptor.

MSAM multiple sequential access method, *IBM*.

MSB 1. *Mass spectrometry bulletin*. Database, Mass Spectrometry Data Centre, Royal Society of Chemistry (UK). 2. most significant bit, data compaction.

MSBR maximum storage bus rate.

MSC main switching centre, telecommunications. 2. mass storage controller, computing. 3. message switching concentration. 4. Microsystems Centre. Set up with *DoI* grant to give advice to business and industry (UK).

MSCU modular store control unit.

MSCW marked stack control word.

MSD 1. modem sharing device. 2. most significant digit.

MSDB main storage database.

MSDOS Microsoft *DOS*, *IBM*.

MSDS message switching data service.

MSEC millisecond. 10^{-3} seconds.

MSF mass storage facility.

MSI medium scale integration. *IC* technology.

MSIO mass storage input/output.

MSK 1. minimal shift keying. 2. Mostek Corporation. Microprocessor manufacturer (US).

MSKM minimum shift keyed modulation.

MSL machine specification language.

MSM memory storage module.

MSMLCS mass service mainline cable systems.

MSNF multi-system networking facility.

MSOS mass storage operating system, *CDC*.

MSP 1. Management System Programmers Ltd. (UK). 2. mass storage processor. 3. medium speed printer. 4. modular system programs, *IBM*.

MSR 1. mark sense reading. 2. mass storage resident. 3. memory select register.

MSS 1. mass storage (sub)system. 2. multi-task single stream system, Raytheon (US).

MSSC mass storage system communication.

MSSS *Mass spectral search system*. Databank, *EPA, NHLI* and *UKCIS*.

MST monolithic systems technology.

MSTG Mass Storage Task Group, *CODASYL*.

MSU 1. main switching unit, telecommunications. 2. mass storage unit.

MSV mass storage volume.

MT 1. machine translation. 2. magnetic tape. 3. Mannesman Tally. Printer manufacturers. 4. mechanical translation. 5. multiple transfer. 6. multi-tasking.

MTA 1. magnetic tape accessory. 2. multiple terminal access.

MTBE mean time between errors.

MTBF mean time between failures.

MTBI mean time between interrupts.

MTBM mean time between maintenance.

MTC 1. magnetic tape cassette. 2. magnetic tape channel. 3. magnetic tape controller. 4. master tape control. 5. message transmission controller.

MTCA multiple terminal communications adaptor.

MTCF mean time to catastrophic failure.

MTCH magnetic tape channel.

MTCS minimum teleprocessing communications system.

MTCU magnetic tape control unit.

MTDR real-time disc monitor, *SEMIS*.

MTE multiple terminator emulator.

MTH magnetic tape handler.

MTM 1. multiple terminal manager, *IBM*. 2. multi-tasking monitor, *SEMIS*. 3. multi-terminal monitor, Interdata.

MTMF multiple task management feature.

MTOS 1. magnetic tape operating system. 2. multi-tasking operating system.

MTP 1. magnetic tape processor. 2. modular terminal processor.

MTPC minimal total processing time.

MTR magnetic tape reader.

MTS 1. message telephone service. 2. microprocessor training system, Integrated Computer Systems (US). 3. million (10^6) transitions per second, of magnetic storage. 4. mobile telephone service. 5. modem test set. 6. multiple terminal system.

MT/ST magnetic tape selectric typewriter, *IBM*.

MTT 1. magnetic tape terminal. 2. magnetic tape transport. 3. message transfer time.

MTTD mean time to diagnosis.

MTTF mean time to failure.

MTTI Maszaki Tudomanyos Tajekoztato Interzetben. Institute of Scientific and Technical Information (Hungary).

MTTR mean time to repair.

MTTS multi-task terminal system.

MTU 1. magnetic tape unit. 2. Manchester terminal unit. *TDM* terminal using Manchester code. 3. memory transfer unit. 4. multiplexer and terminal unit.

M2FM modified modified frequency modulation. Also *MMFM*.

MTX mobile telephone exchange.

MU memory unit.

MUDPIE *Museum and university data program and exchange*. Publication, Smithsonian Institution (US).

MUF maximum usable frequency, signal transmission.

MUG *MARC* Users Group (UK).

MUI module interface unit.

MULS *Minnesota union list of serials*, included in *CONSER*.

MULTICS multiplexed information and computing service, Honeywell (US).

MULTP multiplier.

MUM 1. methodology for unmanned manufacture. Robotics project (Japan). 2. multi-user message. 3. multi-user monitor.

MUMPS Massachusetts General Hospital utility multiprogramming system.

MUMS multiple use *MARC* system. Online retrieval system, *LC*.

MUNIDB *Municipal bonds databank*. Originated and operated by Telestat Systems (US).

MUP multiple utility peripheral.

MUPID multi-purpose universal programmable intelligent decoder.

MUS 1. manual update service, *HP*. 2. multiprogramming utility system, Regnecentralen (Denmark).

MUSA multiple unit steerable array. Antenna.

MUSICOL musical instruction composition oriented language. Computer language for music composition, State University of New York at Buffalo (US).

MUSIL multiprogramming utility system interpretive language, Regnecentralen (Denmark).

MUSS Manchester University software system. Operating system.

MUSTARD museum and university storage and retrieval of data. Smithsonian Institution (US).

MUSTRAN music translation. Music data-processing system, Indiana University (US).

MUTEX multi-user transaction executive, *SEMIS*.

MUX multiplexer.

MV mean value.

MVDS modular video data system, Sperry Univac (US).

MVM minimum virtual memory.

MVP multiple virtual processing.

MVS multiple virtual storage, *IBM*.

MVS/SE *MVS*/system extension.

MVS/SP *MVS*/system product.

MVT multiprogramming with variable number of tasks. Control program, *IBM*.

MVT/TSO *MVT*/time sharing option.

MW 1. meetings word. Searchable field, *BLAISE*. 2. million (10^6) words.

MWD megaword.

MWI message waiting indicator.

MWR magnetic tape write memory.

MWS multi-workstation.

MXE mobile electronic exchange.

MXU mobile exhibition unit.

MYCOS MY compact operating system, Toshiba (Japan).

MYCOS/SS *MYCOS* support system.

N

N 1. nano (10^{-9}). 2. negative.

NA 1. not applicable. 2. numerical analysis.

NAARS *National automated accounting research system*. Database on finance and accountancy (US).

NABD Normen Ausschuss Bibliotheks und Dokumentationswesen. Standards committee on library work and documentation (FRG).

NAC 1. network access controller. 2. Network Advisory Committee, *LC*.

NACIS *National credit information service*. Databank on finance and business, *EIS* and TRW Inc. (US).

NAD node administration.

NAE National Academy of Engineering (US).

NAF network access facility.

NAHB National Association of Home Builders of the United States. Originator and databank on housing and mortgages (US).

NAICC National Association of Independent Computer Companies (US).

NAIPRC Netherlands Automated Information Processing Research Centre.

NAK negative acknowledgement. Data communication.

NAL 1. National Agricultural Library. Database originator (US). 2. new assembly language.

NAM 1. network access machine. 2. network access method, *CDC*.

NAN network application node.

NAND 'not' and 'and'. Compound logical operator.

NANTIS Nottingham and Nottinghamshire technical information service. Library cooperative scheme (UK).

NANWEP Navy numerical weather prediction. Computer system (US).

NAP 1. network access protocol. 2. noise analysis program.

NAPS National Auxiliary Publications Service (US).

NARIC National Rehabilitation Information Center. Originator and databank (US).

NARS National Association of Radiotelephone Systems (US).

NARS-A1 National archive and record service – automation 1. Computerized archive management system, *NARS* (US).

NAS 1. National Academy of Science (US). 2. National Advanced Systems. Computer manufacturer (US).

NASA National Aeronautics and Space Administration (US).

NASA-STAR *NASA scientific and technical reports*. Database, *NASA*.

NASIC Northeast Academic Science Information Center. Information broker and consultancy (US).

NASIS National Association for State Information Systems (US).

NASW National Association of Science Writers (US).

NATCENTATHLIT *Natcentfor athletic literature*. Database, University of Birmingham (UK).

NATIS national information systems, *UNESCO* now (with *UNISIST*) *GIP*.

NAU 1. network addressable unit. 2. network administration utilities, Honeywell.

NAVA National Audio Visual Association (UK).

NAVFAC *Navy faces*. Databank of navy specifications (US).

NAWDEX national water data exchange. US Geological Survey system.

NBA net book agreement. Book pricing agreement (UK).

137

NBCD natural binary-coded decimal.

NBF Norsk Brannvern Forening. Database originator (Norway).

NBFM narrow band frequency modulation.

NBM non-book material.

NBMB N binary digits – M binary digits. Coding technique.

NBMCR non-book materials cataloguing rules. See *LANCET* rules.

NBPM narrow band phase modulation.

NBS National Bureau of Standards (US).

NBS-SIS *NBS* – standard information services.

NC 1. network control. 2. no charge. 3. no circuit.

NCAM network communication access method.

NCB network control block.

NCBE National Clearing House for Bilingual Education. Database originator (US).

NCC 1. National Computing Centre. Major computing resource, originator and operator of database and databanks on computing (UK). 2. network coordination centre.

NCCAN National Center on Child Abuse and Neglect. Originator and database (US).

NCCF network communications control facility.

NCD network cryptographic device.

NCE network connection element.

NCEC National Chemical Emergency Centre, *UKAEA*.

NCECS North Carolina Educational Computing Services (US).

NCES National Center for Educational Statistics. Database originator (US).

NCFR National Council on Family Relations. Originator and databank (US).

NCH network connection handler.

NCHS National Center for Health Statistics. Originator and database (US).

NCI 1. National Cancer Institute. Database originator (US). 2. National Computing Industries (US). 3. non-coded information.

NCJRS *National criminal justice reference service*. Database, US Department of Justice.

NCL network control language.

NCLIS National Commission on Libraries and Information Science (US).

NCM network control module.

NCMH *National clearinghouse for mental health information*. Database, *NIMH*.

NCMS National Classification Management Society (US).

NCMT numerical controlled machine tool.

NCN 1. network control node. 2. Nixdorf Communications Network, Nixdorf (FRG).

NCOLUG North Carolina Online User Group.

NCOM Database, *NORDICOM*.

NCOS non-concurrent operating system, Sperry Univac (US).

NCP network control program, *IBM*.

NCP/VS *NCP* virtual storage.

NCR National Cash Register. Computer manufacturer (US).

NCRDS *National coal resources data system*. Databank, *USGS*.

NCS 1. *National crime survey*. Databank, *DUALabs* (US). 2. *NCR* Century software. 3. network control system. 4. non-conventional system, of post coordinate indexing.

NCSH National Clearinghouse for Smoking and Health. Originator and database (US).

NCSS National Conversational Software Systems Inc. (US).

NC/STRC North Carolina Science and Technology Research Center. *IR* center, University of North Carolina (US).

NCUGAE National Computer User Group in Agricultural Education (UK).

NDB *Numeric database, INPADOC*.

NDBMS network database management system.

NDC 1. national data communication. 2. network diagnostic control.

NDEX *Newspaper index.* Database on content of major US newspapers, Bell and Howell.

NDL network definition language, Burroughs (US).

NDLC network data link control.

NDMS network design and management system.

NDOS new disc operating system, Hitachi (Japan).

NDPS national data processing service. *DP* and transmission service, *BT*.

NDR non-destructive read.

NDRO non-destructive readout.

NDT non-destructive testing.

NE not equal to. Relational operator.

NEANDC Nuclear Energy Agency Nuclear Data Committee (France).

NEAT *NCR* electronic autocoding technique.

NEB National Enterprise Board. Now merged with *NRDC* to form *BTG* (UK).

NEBULA natural electronic business users language.

NEC 1. National Electronics Conference (US). 2. National Electronics Council (UK). 3. Nippon Electric Company (Japan).

NED new editor. Text processing software.

NEDS *National emissions data system.* Databank, *EPA*.

NEEDS *National emergency equipment data system.* Databank, Environment Canada.

NEEP New England Economic Project. Originator and databank.

NEFAX *NEC FAX.* See *NEC (2)* and *FAX*.

NEG negative.

NEI *Nordic energy index.* Databank, Riso Library (Denmark).

NEIBR Norsk Institutt for By-og Regionforskning. Norwegian Institute for Urban and Regional Research. Database originator.

NEISS *National electronic injury surveillance system.* Databank, *CPSC*.

NEL National Engineering Laboratory. Part of *DoI* (UK).

NEL-APPES Databank on physical properties of compounds, *NEL*.

NELINET New England library information network. Cooperative library network (US).

NEMA National Electricity Manufacturers' Association. Originator and databank (US).

NEPHIS nested phrase indexing system. Automated indexing system.

NERC National Environmental Research Council. Originator and database, also grant awarding body (UK).

NERIS *National energy referal information system.* Databank of expertise, *EIC*.

NESS network and evaluation simulation system, *ICL*.

NETMUX network multiplexer, *BT*.

NETSET network systems and evaluation technique.

NETWORK North East Tyneside library cooperative system (UK).

NEUDATA *Neutron data under direct access.* Databank, *OECD*.

NEVA facsimile transceiver (USSR).

NEXIS Newspaper database, Mead Data Control (US).

NF 1. noise figure, telecommunications. 2. non-fragments. Searchable field, *ESA-IRS*.

NFAIS National Federation of Abstracting and Indexing Services (US).

NFAM network file access method.

NFAP network file access protocol.

NFBS Nordiske Forskningsbibliotekens Samarbejdskomite. Research libraries cooperative committee (Scandinavia).

NFE network front end.

NFLCP National Federation of Local Cable Programers, *CATV* (US).

NFPA National Fire Protection Association. Databank originator (US).

NFT networks file transfer.

NGA National Graphical Association. Trade union (UK).

NGP network graphics protocol.

NHELP new Hitachi effective library for programming.

NHK Nippon Hose Koyokai. Japanese National Broadcasting Corporation.

NHLI National Heart and Lung Institute. Databank originator (US).

NHP network host protocol.

NHPC National Historical and Publications Commissions. Database compilers (US).

NHTSA National Highway Traffic Safety Administration. Databank originator (US).

NI 1. Normenausschuss Informationsverarbeitung. Standards committee for information processing (FRG). 2. numerical index.

NIAMDD National Institute of Arthritis, Metabolism and Digestive Diseases. Database originator (US).

NIB 1. negative impedance booster, electronics. 2. node initialization block. 3. *Normeninformationsbank*. Databank on standards, *DIN*.

NIC 1. network interface control. 2. 1900 indexing and cataloguing, *ICL*.

NICC Nationalized Industries Computer Committee (UK).

NICEM *National information center for educational materials*. Database, University of Southern California (US).

NICOL 1900 commercial language, *ICL*.

NICS network integrity control system.

NICSEM *National information center for special educational materials*. Databank, University of Southern California (US).

NIDA numerically integrated differential analyser.

NIDOC National Information and Documentation Centre (Egypt).

NIEHS National Institute of Environmental Health Sciences. Databank originator (US).

NIER National Institute for Educational Research (Japan).

NIF network information files, Burroughs (US).

NIFO next in first out.

NIH not invented here. *DP* colloquialism.

NILUG National Independent Lynx User Group. Minicomputer user group (UK).

NIM 1. network interface machine, Datapac. 2. network interface monitor. 3. newspapers in microform.

NIMH National Institute of Mental Health. Databank and database originator (US).

NIMIS *National instructional materials information system*. Databank, University of Southern California (US).

NIMMS 1900 integrated modular management system, *ICL*.

NIMSCO *NODC index for instrument measured subsurface current observation*. Databank, *NODC*.

NINES Norfolk Information Exchange Scheme (UK).

NIOSH National Institute for Occupational Safety and Health. Databank originator (US).

NIOSHTIC *NIOSH technical information center*. Databank, *NIOSH*.

NIP non-impact printer.

NIPA *National income and product accounts*. Databank, *DOC*.

NIPDOK Nippon Documentesyon Kyokai. Japanese documentation society.

NIPO negative input/positive output.

NIPS *NMCS* information processing system.

NIR next instruction register.

NIRI National Information Research Institute (US).

NIROS Nixdorf real-time operating system.

NIS 1. network information services. 2. network interface system.

NISH *National information sources on the handicapped*. Database, clearing house on the handicapped via *BRS* (US).

NISP national information system for psychology (US).

NISPA national information system for physics and astronomy (US).

NIST national information system for science and technology (Japan).

NIT non-intelligent terminal. Computer hardware.

NITC National Information Transfer Centre, *UNESCO*.

NITCCU Northern Information Technology Centre Consultancy Unit (UK).

NJCL network job control language.

NJE network job entry.

NJI network job interface.

NJIT New Jersey Institute of Technology (US).

NJUS *Netherlands jurisprudence*. Database, Koninklijke Vermande B.V.

NL 1. new line. Control character. 2. non-loaded, telecommunications.

NLA 1. National Library of Australia. Host and database originator. 2. normalized load access.

NLC 1. National Library of Canada. Database originator. 2. new line character. Control character.

NLLST national lending library for science and technology. Replaced by *BLLD*.

NLM National Library of Medicine. Host and database compiler (US).

NLOS natural language operating system.

NLP non-linear programming.

NLR noise load ratio.

NLS 1. National Library of Scotland. 2. online system. System for computer-mediated communication developed at Stanford Research Institute (US).

NLSLS National Library of Scotland lending services.

NM 1. nanometre (10^{-9} metres). 2. network manager.

NMA 1. National Microfilm Association, later National Micrographics Association, now *AIIM*. Trade association (US). 2. National Microform Association (US).

NMAA National Machine Accountants Association (US).

NMC network management centre.

NMCS National Military Command System (US).

NMF new master file.

NMOS negative channel metal oxide semiconductor. Electronics component.

NMR Nuclear magnetic resonance literature retrieval system. Databank, *NIAMDD* and Preston Technical Abstracts Co. (US).

NMRC National Microelectronics Research Centre (Ireland).

NMS 1. *National master specification*. Databank on construction standards and specifications (US). 2. network management services.

NMSD next most significant digit.

NN nearest neighbour. Search term.

NNA new network architecture.

NNC national network congestion signal, telecommunications.

NND national number dialling, telecommunications.

NNI *National newspaper index*. Database, Information Access Corporation (US).

NO notes. Searchable field, *SDC*.

NOBIN Nederlands Orgaan ter Bevordering van de Informatieverzorging. Netherlands organization for the promotion of information provision.

141

NOCI Nederlandse Organisatie voor Chemische Informatie. Netherlands organization for chemical information.

NOCP network operator control program.

NOD Norsk Oseanografisk Datasenter. Databank originator (Norway).

NODAL network-oriented data acquisition language.

NODC National Oceanographic Data Center. Databank originator (US).

NOF *NCR* optical font.

NOHS *National oceanographic hazard survey*. Databank, *NIOSH*.

NOI node operator interface.

NOMA National Office Management Association (US).

NOR 'not' and 'or'. Compound logical operator.

NORC National Opinion Research Center, University of Chicago (US).

NORD Norsk Data. Manufacturer and computer series.

NORDDOK Nordic Committee for Information and Documentation.

NORDFORSK Nordiska Samarbeitsorganisationen for Tekniksnaturvenskaplig Forskning. Scandinavian council for applied research.

NORDICOM Nordic Documentation Centre for Mass Communications Research. Database originator.

NORIANE *Normes et règlements – informations automatisées accessibles en ligne*. Standards databank, *AFNOR*.

NORINDOK Norsk Komite for Informasjon og Dokumentasjon, Norway.

NORIS Norwegian studies in legal informatics.

NORMARC Norwegian *MARC*, Royal University of Norway.

NORMATERM *Normalisation, automatisation de la terminologé*. English-French terminology databank, *AFNOR*.

NOS network operating system, *CDC*.

NOS/BE *NOS*/batch environment.

NOSP network operating support program.

NOTIS Northwestern online totally integrated system. Library automation project, Northwestern University (US).

NOU *Nouvelles*. Database on water resource items in French language newspapers, environment (Canada).

NP 1. name of publisher. Searchable field, Dialog *IRS*. 2. non-patents. Searchable field, *ESA-IRS*. 3. no parity.

NPA 1. National Planning Association. Originator and databank (US). 2. numbering plan area, telecommunications.

NPD network protective device.

NPIS national physics information system, American Institute of Physics.

NPIU network processing and interface unit.

NPL National Physical Laboratory. Databank producer and operator, *DOI*.

NPP network protocol processor.

NPR noise/power ratio.

NPRL non-procedural referencing language.

NPS 1. network processing supervisor, Honeywell. 2. network processor system, *ICL*.

NPSWL new program status word location.

NPT network planning unit.

NPU network processing unit.

NRC 1. National Referal Center, *LC*. 2. National Research Council of Canada. Also *NRCC*.

NRCC National Research Council of Canada. Also *NRC*.

NRCCL Norwegian Research Centre for Computers and Law.

NRCd National Reprographic Centre for Documentation Study, evaluation and information body (UK).

NRCMF *NRC master file*. Databank, *LC*.

NRCST National Referal Center for Science and Technology, *LC*.

NRDC National Research Development Corporation. Now merged with *NEB* to form *BTG* (UK).

NRDF non-recursive digital file.

NRFD not ready for data.

NRI Nomura Research Institute. Databank originator (Japan).

NRL network restructuring language.

NRLSI National Reference Library for Science and Invention. Now British Library: Science Reference Library (UK).

NRM normal response mode.

NRMM national register of microform masters, *LC*.

NRZ non-return to zero, data transmission.

NRZI 1. *NRZ* indicator. 2. *NRZ* inverted. Recording method.

NS nanosecond (10^{-9} seconds).

NSA *Nuclear science abstracts*. Database, Energy Research and Development Administration and *ORNL* (US).

NSC 1. National Semiconductor Corporation (US). 2. network switching centre. 3. nodal switching centre.

NSEC *NS*.

NSF National Science Foundation. Grant awarding body and database originator (US).

NSI 1. next sequential instruction. 2. Norsk Senter for Informatikk. Norwegian Centre for Informatics. Information broker, host and database originator.

NSIC Nuclear Safety Information Center, Atomic Energy Commission (US).

NSIDK Nederlandse Stichting Informatie en Dokumentatiecentrum voor die Kartografie. Netherlands Foundation for Information and Documentation in Cartography.

NSM network security module.

NSP 1. network services protocol, *DEC*. 2. network support processor.

NSR *Nuclear structure references*. Database, *ORNL*.

NSRDS national standard reference data system (US).

NSS 1. network supervisor system. 2. network synchronization subsystem, telecommunications.

NSSDC National Space Science Data Center, *NASA*.

NSTIC Naval Science and Technology Information Centre. Now Defence Research Information Centre (UK).

NSU network service unit, telecommunications.

NTA National Telecommunications Agency (US).

NTAG Network Technical Architecture Group of the Network Development Office, *LC*.

NTC 1. National Telecommunications Conference, *IEEE*. 2. National Translation Center, *SLA* and University of Chicago (US).

NTE network terminating equipment.

NTF no trouble found.

NTH Norges Tekniske Hogskole. Norwegian Institute of Technology. Databank originator.

NTI noise transmission impairment, telecommunications.

NTIA National Telecommunications and Information Agency, Department of Commerce (US).

NTIAC *Non-destructive testing information analysis centre*. Databank, *SWRI*.

NTICL National Technical Information Centre and Library (Hungary).

NTIS 1. National Technical Information Service. Originates database on reports of government agencies (US). 2. *NEC*-Toshiba Information Systems Inc.

NTL 1. Northern Telecom Ltd. (Canada). 2. *Novosti tehniskoi literatury*. Current awareness bulletins, Central Science and Technical Library (USSR).

NTP 1. network terminal protocol.
2. network termination processor.

NTR next task register.

NTSC 1. national television standard code. For broadcasts, videodiscs, videocassettes etc. (US). 2. National Television System Committee. Specification body (US).

NTSK Nordiska Tele-Satelit Kommitton (Norway).

NTT Nippon Telegraph and Telephone. Telecommunications and videotex company (Japan).

NTU network terminating unit.

NU number unobtainable (tone) telecommunications.

NUA network user address. User code for message direction.

NUC *National union catalog* (US).

NUC/codes *NUC codes*. Databank of names and addresses of libraries cited in *CASSI*.

NUCOM *National union catalogue of monographs* (Australia).

NUI network user identifier. Password.

NUL null. Control character.

NUM numeric.

NUMAC Northumbrian universities multiple access computer.

NVRAM non-volatile random access memory.

NVT network virtual terminal.

NWD network wide directory.

NYPL New York Public Library (US).

NYSILL New York State Inter-Library Loans System.

NYTIS New York Times Information Service. Database originator and host.

NZT non-zero transfer.

O operand.

OA 1. office automation. 2. operand address.

O/A on application, usually of prices.

OAAU orthogonal array arithmetic unit.

OAF origin address field.

OAG *Official airline guide.* Database, I. P. Sharp.

OAND *Origin and destination.* Passenger surveys database.

OAP orthogonal array processor. Type of computer.

OAQS online associative query system. *IR* search strategy technique, University of Illinois (US).

OARS office automation reporting service.

OAS office automation system.

OASIS oceanic and atmospheric scientific information system. National Oceanic and Atmospheric Administration (US).

OASYS office automation system.

OATS original article tearsheet service. Current awareness service, *ISL*.

OAU operator assistance unit, telecommunications.

OB 1. octal to binary. 2. output buffer. 3. output bus.

OBAR *Ohio bar.* Legal information service (US).

OBCE Office Belge du Commerce Extérieur. Originator and databank on trade (Belgium).

OBCR optical bar code reader.

OBH office busy hour, telecommunications.

OBNA only but not all.

OC 1. *OCLC* number. 2. old category code. Searchable fields, Dialog *IRS*. 3. *Online chronicle.* Electronic journal on electronic publishing. 4. operating characteristics. 5. operator command.

OCB outgoing calls barred, telecommunications.

OCC 1. Osborne Computer Corporation. Microcomputer manufacturer. 2. other common carrier, telecommunications.

OCCS Office of Computer and Communication Systems, *NLM*.

OCG optimal code generation.

OCH Office for Communication in the Humanities, Leicester University (UK).

OCI 1. Office of Computer Information. Originator and database (US). 2. optically coupled isolator.

OCL 1. operator control language. 2. operation control language. *JCLs.* 3. overall connection loss.

OCLC Online Computer Library Center, formerly Ohio College Library Center. Provider of online services to libraries (US).

OCM oscillator and clock module.

OCP 1. order code processor. 2. output control program. 3. Oxford concordance project, Oxford University (UK).

OCPDB *Organic chemical producers data base.* Databank, *EPA*.

OCR 1. optical character reader. 2. optical character recognition. 3. output control register.

OCR-A *OCRS* typeface.

OCR-B *OCRS* typeface.

OCRS optical character recognition system.

OCS 1. office communications system. 2. operations control system. 3. Oriel Computer Services Ltd. (UK). 4. output control subsystem. 5. overload control subsystem.

OCSL Oriel Computer Services Limited UK).

OCT operational cycle time.

OD 1. octal to decimal. 2. output disable.

ODA operational design and analysis.

ODB 1. output data buffer. 2. output to display buffer, computing.

ODC output data control.

ODD 1. operator distance dialling. 2. optical data digitizer. 3. optical data disc. 4. optical digital disc.

ODE online data entry. User facility for file creation, *ESA-IRS*.

ODESY online data entry system, Burroughs (US).

ODF *Original data file*. Databank, *DAEDAC*.

ODG offline data generator.

ODIN Online Dokumentations und Informationsnetz. *IRS* (FRG).

ODIN PAN scientific information documentation centre of Polish Academy of Sciences.

ODP optical data processing.

ODR output definition register.

ODT 1. octal debugging technique. 2. online debugging technique.

ODU output display unit.

OE output enable.

OEAP operational error analysis program.

OECD Organization for Economic Cooperation and Development. Inter-governmental organization, originates and operates databanks.

OECD/MEI *OECD main economics indicators*. Databank, *OECD*.

OECD/NIA *OECD national income accounts*. Databank, *OECD*.

OEM 1. Office and Electronic Machines (UK). 2. original equipment manufacturer.

OFIS office information system.

OFIX Office of the Future Information Exchange. User group (UK).

OFT optical fibre tube.

OFTEL Office of Telecommunications (UK).

OFTS optical fibre transmission system.

ÖGDI Österreichische Gesellschaft für Dokumentation und Information. Austrian Society for Documentation and Information.

OH octal to hexadecimal.

ÖHB *Österreichische historische Bibliographie*. Database, *IMD*.

OHIONET Ohio network. Library cooperative (US).

OHM-TADS *Oil and hazardous materials – technical assistance data system*. Databank, *EPA*.

OHS Occupational Health Services Inc. Medical databank original (US).

OI *Olje indeks*. Database on oil, *NSI*.

OIB operation instruction block.

ÖIBF Österreichische Institut für Bibliotheksforschung Dokumentations und Informationswesen. Austrian Institute for Librarianship, Documentation and Information Science.

OICP Office of International Communications Policy, *ICA*.

OIDI optically isolated digital input.

OIS office information system.

ÖIZ Österreichisches INISzentrum. Information broker and *INIS* operator (Austria).

OL 1. online. 2. open loop. 3. output latch.

OLAS online acquisitions system. Book acquisition system, Brodart Company (US).

OLB online batch.

OLDB online database.

OLIFLM online image forming light modulator.

OLLT Office of Libraries and Learning Technologies, Department of Education (US).

OLM online monitor.

OLOE online order entry.

OLP online programming.

OLPARS online pattern analysis and recognition system.

OLPS online programming system.

OLRT online real time.

OLS online search.

OL'SAM online search assistance machine, Franklin Institute (US).

OLSUS online system use statistics, *OCLC*.

OLT online test.

OLTE online test, Fujitsu (Japan).

OLTEP online test executive program, *IBM*.

OLTS online test section.

OLTT online terminal test.

OLUC *Online union catalog*. Database, *OCLC*.

OM 1. operations manual. 2. output module.

OMA operations monitor alarm.

OMAC online manufacturing control, *ICL*.

OMAP object module assembly program.

OMD open macro definition.

OMF old master file.

OMKDK Orszagos Muszaki Konyvtar es Documentacios Kozpont. Hungarian Central Technical and Documentation Centre. Host.

OMP optical mark printer.

OMPR optical mark page reader.

OMR optical mark recognition.

ON 1. officer's name. Searchable field, Dialog *IRS*. 2. order number. Searchable field, *SDC*.

ONI operator number identification.

ONT one-time carbon. Computer stationery.

ONTAP online training and practice. Training files for *CA search, Chemname* and *ERIC* from Lockheed (US).

ONTYME-II On Tymenet. Electronic mail service, Tymenet Inc. (US).

OOLUG Oklahoma On Line Users Group (US).

OOPS offline operating simulator.

OOUG Oregon Online User Group (US).

OP 1. operation. 2. out of print.

OPA operator priority access, telecommunications.

OPAL operational performance analysis language.

OP-AMP operational amplifier.

OPC 1. online plotter controller. 2. operation code.

OPCODE operation code.

OPE optimized processing element.

OPER 1. operation. 2. operator.

OPL Official Publications Library, *BL*.

OPM operations per minute. Computer performance measure.

OPOL optimization-oriented language.

OPP octal print punch.

OPR optical pattern recognition.

OPRIS Ohio project for research in information service (US).

OPS 1. operations. 2. operations per second.

OPSKS optimum phase shift keyed signals, telecommunications.

OPSWL old program status word location.

OPSYS operating system.

OPT option.

OPTS online peripheral test system.

147

OPUR object program utility routines.

OPUS Organisation of Professional Users of Statistics (UK).

OQL online query language.

O-QPSK offset *QPSK*.

OR operational research.

ORACLE optical recognition of announcements by coded line electronics. Teletext system (UK).

ORBIT 1. online real-time branch information. Computer programming language. 2. online retrieval of bibliographic information timeshared. *IR* language and software, *SDC*. 3. *ORACLE* binary internal translator. Used with *ORACLE* computer.

ORC orthogonal row computer.

ORCHIS Oak Ridge computerized hierarchical information system, Atomic Energy Commission (US).

ORC youth *Opinion Research Corporation youth*. Databank on teenage consumer and labour market, *ADL*.

ORE overall reference equivalent, telecommunications.

ORICAT original cataloguing system. Non-English catalogue system, *HOBITS*.

ORION online retrieval of information over a network.

ORNL Oak Ridge National Laboratory, Atomic Energy Commission. Host and database originator (US).

ORSA Operations Research Society of America.

ORSTOM Office de la Recherche Scientifique et Technique Outre-mer. Databank originator and operator (France).

ORT operational readiness test.

OS 1. office system. 2. operating system. 3. organizational source. Searchable field, *SDC*.

OSAIS *Oil spillage analytical information service*. Database, Laboratory of the Government Chemist (UK).

OSAM overflow sequential access method.

OSB operational status bit.

OSCL operating system control language.

OSCO organizational source code. Searchable field, *SDC*.

OSCRL operating system command and response language.

OSD 1. online system drivers, *NCR*. 2. optical scanning device.

OSHB one sided height balanced, telecommunications.

OSI 1. open system interconnection. Networking technique. 2. operating system interface.

OSIL operating system implementation language.

OSIRIS online search information retrieval information storage, US Navy.

OSL operating system language.

OSM operating system monitor.

OS/MFT operating system/multiprogramming fixed task, *IBM*.

OS/MIT operating system/multiprogramming variable task, *IBM*.

OSN output sequence number.

OSR 1. operand storage register. 2. optical scanning recognition.

OSS 1. operating system supervisor. 2. operating system support.

OSSL operating system simulation language.

OSTI Office for Scientific and Technical Information. Now *BLR&DD*.

OSUL Ohio State University libraries (US).

OT 1. Office of Telecommunications (US). 2. output terminal.

OTA Office of Technology Assessment (US).

OTAF Data Base *Office of Technology Assessment and Forecasts data base*, US Patent and Trademark Office.

OTC 1. originating toll centre, telecommunications. 2. overseas telecommunications commission.

OTE 1. operational test and evaluation. 2. Authority for Telecommunications (Greece).

OTF optical transfer function.

OTG option table generator.

OTL online task loader.

OTONI Division of theoretical foundations of scientific information, *VINITI*.

OTP Office of Telecommunications Policy (US).

OTRAC oscillogram trace reader, Non-Linear Systems Inc. (US).

OTS 1. orbital test satellite. Communications satellite programme, *ESA*. 2. Oxford text system. Text editing and setting system, *OUP*.

OU 1. object unit. 2. operation unit.

OULCS Ontario University libraries cooperative system (Canada).

OUP Oxford University Press.

OUTRAN output translator, *IBM*.

OVD optical video disc.

OVID online visual display unit interrogation of databases, British Steel.

OWF optimum working facility.

OWL online without limits.

149

P

P 1. permutation. 2. peta (10^{15}). 3. pico (10^{-12}). 4. positive. 5. power. 6. program.

PA 1. paper advance. 2. patent assignee. Searchable field, Dialog and *SDC*. 3. patent countries. Searchable field, *SDC*. 4. patents. Searchable field, *ESA-IRS*. 5. *Physics abstracts*. Database, *INSPEC*. 6. power amplifier. 7. program access. 8. program address. 9. program amount. Searchable field, Dialog *IRS*. 10. Publishers Association (UK). 11. pulse amplifier. 12. purpose and activities. Searchable field, Dialog *IRS*.

PABX private automatic branch exchange, telecommunications.

PAC 1. package assembly circuit. 2. peripheral autonomous control. 3. personal analog computer. 4. polled access circuit. 5. program assembly card.

PACC product administration and contract control. Management computer system, Sperry Rand (US).

PACE packaged *CRAM* executive, *NCR*.

PACT 1. Pandick computerized typesetting. Word processor service (US). 2. programmable asynchronous clustered teleprocessing.

PACTEL Planning Associates for Computers and Telecommunications. Consultancy (UK).

PACX private automatic computer exchange.

PAD 1. packet assembly disassembly. Packet switching technique. 2. positioning arm disc.

PADIA patrol diagnosis, Fujitsu (Japan).

PADL parts and design language. *CAD* software, Rochester University (US).

PADLA programmable asynchronous dual line adaptor.

PADS performance analysis display system.

PAF 1. page address field. 2. peripheral address field.

PAFC phase locked automatic frequency control, telecommunications.

PAGES program affinity grouping and evaluation system.

PAIS Public Affairs Information Service. Database originator (US).

PAIS FLI *PAIS foreign language index*. Database on public administration in languages other than English, *PAIS*.

PAL 1. partition allocation utility, *GRI* Computer. 2. pedagogic algorithmic language. 3. phase alternate line, *TV* standard. 5. programmed array logic.

PALAPA communication satellite system (Indonesia).

PAL-D delay line *PAL*. *TV* decoding system.

PALINET Pennsylvania Area Library Network. Library cooperative (US).

PALM Philips automated laboratory management system. Integrates laboratory control with information management and automated office functions, Philips (Netherlands).

PAL-S simple *PAL*. *TV* decoding system.

PAM 1. peripheral adaptor module. 2. primary access method, Sperry Univac. 3. process automation monitor, *TI*. 4. pulse amplitude modulation.

PAM/D process automation monitor/disc version, *TI*.

PAN polled access network.

PANDA 1. *PRESTEL* advanced network design architecture. 2. programmers' analysis 'n' development aid, Digico (UK).

PANSDOC Pakistan National Scientific and Technology Documentation Centre.

PANTHEON public access by new technology to highly elaborate online networks. Ironic reference to networks named for

PANTS public acceptance of new technologies. Research project, *SSRC* (UK). Mythological gods, eg *ARTEMIS, HERMES* etc.

PAP phase advance pulse.

PAPERCHEM *Paper chemistry*. Database, Institute of Paper Chemistry (US).

PAPTC paper tape controller.

PAR 1. page address register. 2. program address register. 3. program appraisal and review.

PARC Palo Alto Research Center. Computer and *IT* research centre, Xerox (US).

PARCS *Pesticide analysis retrieval and control system*. Databank, *EPA*.

PARIF program for automation retrieval improvement by feedback, *EURATOM*.

PAS 1. *Patent applicant service, INPADOC*. 2. public address system.

PASCAL 1. computer programming language. 2. *Program appliqué à la sélection et la compilation automatique de la littérature*. Database, *CNRS*.

PASLA programmable asynchronous line adaptor.

PASLIB Pakistan Association of Special Libraries.

PASS 1. program aid software system. 2. programmed access security system.

PASSIM President's Advisory Staff on Scientific Information Management (US).

PAST process accessible segment table.

PAT 1. peripheral allocation table, computing. 2. personalized array translator. 3. programmer aptitude test.

PATBX private automatic telex branch exchange, telecommunications.

PATCA phase lock automatic tuned circuit adjustment, telecommunications.

PATLAW *Patent law*. Database of patent law decisions, Bureau of National Affairs (US).

PATOLIS patent online information system. *IRS, JAPATIC*.

PATRICIA practical algorithm to receive information coded in alphanumeric. Information retrieval technique.

PATSEARCH *Patent search system*. Database, Pergamon (UK).

PATSY programmers' automatic testing system.

PATX private automatic telex exchange, telecommunications.

PAU pattern articulation unit, computing.

PAX 1. physical address extension. 2. private automatic exchange, telecommunications.

PB 1. page buffer. 2. peripheral buffer. 3. plug board. 4. proportional band. 5. publications. Searchable field, Dialog *IRS*. 6. publisher. Searchable field, *ESA-IRS* and Dialog.

PBDG push button data generator.

PBN physical block number.

PBS picture building system, *IBM*.

PBU push button unit.

PBX private branch exchange, telecommunications.

PC 1. patent classification. Searchable field, *ESA-IRS* and *SDC*. 2. patentee/company code. Searchable field, *SDC*. 3. path controller. 4. peripheral controller. 5. personal computer. 6. phase code. Searchable field, Dialog *IRS*. 7. processor controller. 8. product code. Searchable field, Dialog *IRS*. 9. program controller. 10. program coordination. 11. program counter. 12. punch card.

PCA 1. physical configuration audit. 2. printed communications adaptor. 3. programmable communication adaptor. 4. protective connecting arrangement. 5. pulse code adaptor. 6. pulse counter adaptor.

PCAM partitioned content addressable memory.

PCB 1. page control block. 2. printed circuit board. 3. process control block. 4. processor command bus. 5. program communication block. 6. program control block.

PCC 1. peripheral control computer. 2. personal code calling, telecommunications.

3. primary category code. Searchable field, *SDC*. 4. program control counter.

PCCS processor common communications system.

PCE peripheral controller enclosure.

PCF program control facility.

PCG programmable character generator.

PCHIS *Population clearing house and information system*. Database, *ESCAP*.

PCI 1. Packet Communications Inc. (US). 2. peripheral controller interface. 3. process control interface. 4. program controlled interruption. 5. programmable communication interface. 6. Protocol Computers Inc. (US).

PCIOS processor common input/output system.

PCIS personal computer information service.

PCjr personal computer—junior, *IBM*.

PCK processor controlled keying, data entry.

PCL 1. parallel communications link. 2. print control language.

PCLA process control language, *TI*.

PCLR parallel communications link receiver.

PCM 1. plug compatible mainframe. 2. plug compatible manufacturer. 3. plug compatible memory. 4. pulse code modulation. Data transmission. 5. punch card machine.

PCMI photo-chromic micro-image. Type of microfiche.

PCMM plug compatible mainframe manufacturer.

PCN *Personal computer news*. Journal (UK).

PCOS 1. primary communications oriented system. 2. process control operating system.

PCP 1. parallel cascade processor. 2. primary control program. 3. primary cross-connection point. 4. programmable communication processor.

PCR 1. page control register. 2. print command register. 3. program change request.

PCRC Primary Communications Research Centre, University of Leicester (UK).

PCS 1. personal computing system. 2. plastic coated silica, for fibre optic systems. 3. programmable communications subsystem. 4. punch card system.

PCSFSK phase comparison sinusoidal frequency shift keying.

PCT 1. Patent Cooperation Treaty. 2. peripheral control terminal.

PCTIS Preston Commercial and Technical Information Service (UK).

PCU 1. peripheral control unit. 2. processor control unit. 3. program control unit. 4. punched card unit.

PCW program control word.

PD 1. plasma display. 2. process data. Searchable field, Dialog *IRS*. 3. publication data. Searchable field, Dialog, *NLM* and *SDC*.

PDA physical device address.

PDC 1. parallel data controller. 2. Philosophy Documentation Center. Database originator (US). 3. *Photonuclear data center*. Databank, *NBS*. 4. *Predecessors and defunct companies*. Database, Financial Post of Canada via *QL* Systems.

PDD post dialling delay, telecommunications.

PDED partial double error detecting.

PDF 1. processor defined function. 2. *Power diffraction file*. Databank, *JCPDS*.

PDI picture description instruction. Videotex page format.

PDIN Pusat Dokumentasi Ilmiah Nasional. National Science Documentation Centre (Indonesia).

PDIO parallel digital input/output.

PDL 1. procedure definition language. 2. program design language. 3. programmable data language. 4. Publishers' Databases Ltd. Electronic publishing consortium (UK).

PDM 1. practical data manager, Hitachi (Japan). 2. pulse duration modulation, data transmission.

PDN public data network.

153

PDP 1. plasma display panel. 2. procedure definition processor. 3. programmed data processor.

PDPS parts data processing system. Bell Telephone system.

PDR 1. page data register. 2. Pharma Documentation Ring. Pharmaceutical companies' documentation cooperative. 3. processing data rate.

PDS 1. periodicals data system. Cataloguing system, *MINICS/PDS*. 2. personal data system, *ICL*. 3. *Petroleum data system*. Databank, University of Oklahoma (US). 4. photo-digital store. 5. private database service. Software for inhouse use, *BRS*.

PDT 1. physical device table, for peripherals status. 2. programmable data terminal, *DEC*.

PDU programmable delay unit.

PDX private digital exchange, telecommunications.

PE 1. parity error. 2. phase encoding, for magnetic tape recording. 3. processing element, of *CPU*.

PEACESAT Pan-Pacific editing and communication experiment by satellite, University of Hawaii (US).

PEARL 1. periodical enquiry, acquisition and registration locally. Periodical acquisition software, Blackwells (UK). 2. periodicals automation Rand Library, Rand Corporation (US). 3. process, experiment and automation real-time language. 4. programmed editor and automated resources for learning.

PEAS Pacific's electronics acquisition service, Pacific University (US).

PEBUQUILL prêt entre bibliothèques des Universités de Québec. University interlibrary loans (Canada).

PEM 1. photo-electromagnetic. 2. process execution module. 3. processing element memory.

PEMS *Pesticide enforcement management system*. Databank, *EPA*.

PENCIL pictorial encoding language.

P/E News *Petroleum energy business news index*. Database, *API*.

PEP 1. partitioned emulation program. 2. peak envelope power, telecommunications.

PEPE parallel element processing ensemble, Burroughs/US Army.

PER 1. program event recording. 2. program execution request.

PERA system Project Engineering Research Association system. Component classification code (UK).

PERT programme evaluation and review technique. Management technique, also *CPM*.

PES 1. photo-electric scanner. 2. program execution system.

PESTAB *Pesticides abstracts*. Database, *EPA*.

PESTDOC *Pesticide documentation*. Database, Derwent Publications (UK).

PET 1. peripheral equipment tester. 2. personal electronic transaction computer, *CBM*. 3. program evaluator and tester.

PETROEX *Petroleum products exchange data clearing house*. Databank, *GEISCO*.

PEU port expander unit.

PF 1. page footing. 2. picofarad (10^{-12} farads). 3. power factor. 4. programmable function. 5. pulse frequency. 6. punch off.

PFA program and file analysis.

PFEP programmable front end processor.

PFI pack file indexer, *CDC*.

PFK programmed function key.

PFM pulse frequency modulation.

PFP program file processor.

PF/R power fail/restart.

PG pulse generator.

PGA programmable gate array.

PGI General Information Programme, *UNESCO*.

PGP programmable graphics processor.

PH 1. page heading. 2. phase.

PHD parallel head disc.

PHI *Philosophie informationssystem.* Database, Universität Düsseldorf (FRG).

PHILQA Philips question answering system. Natural language research project, Philips Research Laboratories (Netherlands).

PHILSOM periodicals holdings in the Library of the School of Medicine, Washington University, St. Louis (US).

PHLAG Philips load and go. Programming system.

PHOTO *Bottom photograph camera selection file.* Databank, *NODC*.

PI 1. periodical index term. Searchable field, Dialog *IRS*. 2. *Philosophers' index.* Database, Dialog (US). 3. priority interrupt. 4. process image. 5. processor interface. 6. program interrupt. 7. programmed instruction.

PIA 1. peripheral interface adaptor. 2. Printing Industries Association of America.

PIB programmable input buffer.

PIC 1. position independent code. 2. priority interrupt controller. 3. process interface control. 4. processor interconnection channel. 5. programmable interrupt controller. 6. publishers' information card, now *IBIS* service.

PICA Project for integrated catalogue automation. Cataloguing cooperative, Royal Netherlands Library.

PICU parallel instruction control unit.

PID 1. proportional integral derivative. 2. pseudo interrupt device.

PIDCOM process instruments digital communication system, Beckman Industries (US).

PIDS *Parameter inventory display system.* Databank, *NODC*.

PIE 1. *Pacific islands ecosystems.* Database, *FWS*. 2. parallel instruction execution. 3. parallel interface element.

PIIC Pergamon International Information Corporation. Operates Pergamon Infoline.

PIM 1. pilot machine, *CIIHB*. 2. processor interface module.

PIN 1. personal identification name. 2. personal identification number. 3. piece identification number. 4. Private Intelligent Networker. Data network, Gandalf. 5. program identification number.

PINCCA *Price index numbers for current cost accounting.* Databank, *SIA*.

PINT priority interrupt controller.

PIO 1. parallel input/output. 2. peripheral input/output. 3. photocomposition input option, Wang (US). 4. programmed input/output.

PIOCS 1. parallel input/output control system. 2. physical input/output control system.

PIOU parallel input/output unit.

PIP 1. path independent protocol. 2. peripheral interchange program. 3. *Pollution information project.* Database, *NRC* (Canada).

PIPS pattern information processing system.

PIQ parallel instruction queue.

PIR program incident report.

PIRA Paper Printing and Packaging Industries Research Association. Research organization and its database on paper printing and packaging (UK).

PIRETS Pittsburgh retrieval system. *IR* software, University of Pittsburgh (US).

PIRS 1. personal information retrieval system. 2. *Philosopher's index retrieval system.* Database, *PDC*.

PIS *Patent inventor service, INPADOC.*

PISAL *Periodicals in South African libraries.* Database, *CSIR*.

PISW program interrupt status word.

PIT 1. Printing and Information Technology Division, *PIRA*. 2. protection identification key.

PITCOM Parliamentary Information Technology Committee (UK).

PIU process interface unit.

PIW program interrupt word.

PJ 1. picojoule (10^{-12} joules). Measure of logic gate efficiency. 2. project name. Searchable field, Pergamon Infoline.

PKD programmable keyboard display.

PL 1. patent location. Searchable field, *SDC*. 2. place of publication. Searchable field, *NLM*. 3. private line. 4. procedural language. 5. public library.

PLA 1. print load analyser. 2. programmable line adaptor. 3. programmable logic array. 4. programmed logic array.

PLACE programming language for automatic checkout equipment.

PLAN 1. program language analyser. 2. programming language 1900, *ICL*.

PLANES programmed language enquiry system.

PLANET private local area network, Racal (UK).

PLAS program logical address space.

PLATO programmed logic for automated teaching operations, University of Illinois (US).

PLC 1. programmable line controller, Nixdorf (FRG). 2. programmable logic control. 3. Programming Language Committee, *CODASYL*.

PLCA parallel line communication adaptor.

PLD partial line down. Control character.

PLF page length field.

PLI private line interface.

PLJ permanent loop junctor.

PLL phase locked loop.

PLM passive line monitor, Datapoint (US).

PL/M programming language for microprocessors.

PLO phase locked oscillator.

PL/1 programming language/1. Computer programming language.

PLP 1. presentation level protocol. Information presentation standard. 2. procedural language processor.

PLPA pageable link-pack area.

PLR program lock-in register.

PLS 1. physical signalling. 2. Publishers Licensing Society. Joint *PA, PPA* and *ALPSP* body (UK).

PLTTY private line teletypewriter service.

PLU partial line up. Control character.

PLUS 1. program language for user's system, Hitachi (Japan). 2. program library update system.

PLZ programming language Zilog (US).

PM 1. phase modulation, data transmission. 2. preventive maintenance. 3. processor module.

PMA 1. physical medium attachment, of *MAU*. 2. physical memory address. 3. Prime macro-assembler, Prime Computers (US). 4. priority memory access. 5. priority memory address. 6. protected memory address.

PMAR page map address register.

PMB *PROM* memory board.

PMB (Canadian) Canadian Print Measurement Bureau. Originator and its databank on consumer behaviour.

PMBX private manual branch exchange. Telecommunications.

PMD program module dictionary.

PME processor memory enhancement.

PMEST personality, matter, energy, space, time. Colon classification facets.

PMI Precision Monolithics Inc. *IC* manufacturer (US).

PMIC parallel multiple incremental computer.

PML physical memory level.

PMLC programmed multi-line controller.

PMM programmable microcomputer module.

PMMb parallel memory to memory bus.

P-MOS positive channel metal oxide semiconductor.

PMS 1. performance management system. 2. Picturephone Meeting Service. Teleconferencing service, *AT & T*. 3. processor memory switch. 4. project management system, *IBM*. 5. public message service, Western Union (US).

PMSX processor memory switch matrix.

PMX 1. packet multiplexer. 2. protected message exchange.

PN 1. patent number. Searchable field, various *OLS*. 2. personal name. Searchable field, *BLAISE*. 3. phase name. Searchable field, Dialog *IRS*. 4. place of publication class number. Searchable field, *NLM*. 5. product name. Searchable field, Dialog *IRS*. 6. project number. Searchable field, Dialog and *SDC*. 7. programmable network. 8. project number. Searchable field, Dialog and *SDC*. 9. punch on.

PNA packet network adaptor, linking *PSS* and PSTN systems, telecommunications.

PNAF potential network access facility.

PNBC Pacific Northwest Bibliographic Center. Library cooperative (US and Canada).

PNC 1. police national computer (UK). 2. *PRIMENET* node controller.

PNCC partial network control centre.

PNI *Pharmaceutical news index*. Database, Data Courier Inc. (US).

PNP P-type, N-type, P-type transistor.

PNSC packet network service centre, telecommunications.

PO 1. parity odd. 2. performing organization. Searchable field, Dialog *IRS*. 3. post office (UK). 4. pulse output.

POC process operator console.

POCS Patent Office classification system (US).

PODA priority oriented demand assignment.

POETRI Programme on Exchange and Transfer of Information, *UN*.

POEU Post Office Engineering Union. Trade union (UK).

POF point of failure.

POGO programmer oriented graphic operation.

POL 1. program oriented language, Burroughs (US). 2. program oriented language. Type of high level language.

POLANG polarization angle, telecommunications.

POLGEN problem oriented language generator.

POLLS Parliamentary online library study, *UKAEA*.

POLYDOC polytechnical documentation. Computer system, Scandinavia.

POM printout microfilm

POMSA Post Office Management Staff Association. Now *CMA*.

POPINFORM *Computerized population information*. Databank, Center for Population and Family Health (US).

POPLINE *Population online*. Databank, Johns Hopkins, Columbia and Princeton Universities (US).

POPS process operating system, Toshiba (Japan).

POPSI postulate-based permuted subject index.

POR problem oriented routine.

POS 1. point of sale. 2. primary operating system. 3. professional operating system, for microcomputers, *DEC*.

POSH permuted on subject headings. Indexing technique.

POST *Polymer science and technology*. Database, *CAS*.

POT potentiometer.

POTS plain old telephone service.

POUNC Post Office Users' National Council (UK).

POWU Post Office work unit. Computer performance measure, *BT*.

PP 1. parallel processor. 2. peripheral processor. 3. place of publisher. Searchable field, Dialog *IRS*. 4. print positions. 5. program product, *IBM*.

PPA Periodical Publishers Association (UK).

PPB *PROM* programmer board.

PPC 1. print position counter. 2. pro-personal computer, Ferranti Argus.

PPD pulse type phase detector.

PP-DC programming panels and decoding circuits.

PPDS *Physical property data service*. Databank, Institution of Chemical Engineers (UK).

PPE problem program evaluator.

PPI programmable peripheral interface.

PPIB programmable protocol interface board.

PPIU programmable peripheral interface unit.

PPL polymorphic programming language.

PPM 1. periodic pulse metering. 2. previous processor mode. 3. pulse position modulation, data transmission.

PPP parallel pattern processor.

PPQA pageable partition queue area, computing.

PPS 1. page printing system, Honeywell (US). 2. page processing system, Honeywell (US). 3. parallel processing system. 4. programmed processor system. 5. pulses per second, data transmission.

PPSC 1. Privacy Protection Study Commission (US). 2. processor program state control.

PPSS public packet switching service (UK).

PPT punched paper tape.

PPU 1. peripheral processing unit. 2. pre-processor utility.

PPX 1. packet protocol extension. 2. private packet exchange.

PQA protected queue area.

PR 1. priority country. Searchable field, Pergamon Infoline and *SDC*. 2. progress report. Searchable field, Dialog *IRS*.

PRA *Peace research abstracts*. Database, *IPRA*.

PRBS pseudo random binary sequence.

PRC 1. Postal Rate Commission (US). 2. procession register clock. 3. programmed rate control.

PRECIS 1. pre-coordinate indexing system. 2. preserved context index system, *BNB*.

PRE-MED *Previous to appearance in MEDLINE*. Database of current clinical medicine, *BRS*.

PRESTEL viewdata system, British Telecom. Not an acronym.

PREXTEND PRESTEL extended. Advanced *PRESTEL*.

PRF 1. potential requirements file. 2. pulse repetition frequency.

PRI pulse repetition interval.

PRIMENET Prime network software package, Prime Computers (US).

PRIMORDIAL primary order dial. Automated document ordering facility, *ESA-IRS*.

PRIMOS Prime operating system, Prime Computers (US).

PRINCE Parts Reliability Information Center, *NASA*.

PRISM 1. personnel record information system for management, civil service (UK). 2. programmed integrated system maintenance.

PRK phase reversal keying.

PRLC Pittsburgh Regional Library Center. Library cooperative (US).

PRNET packet radio network.

PRO 1. programmable remote operation. 2. Public Record Office (UK).

PROCOMP program compiler.

PROFIT program for financed insurance techniques.

PROLOG programming in logic. Advanced programming language.

PROM programmable read-only memory. Computer memory.

PROMIS 1. problem oriented medical information system. 2. project oriented management information system.

PROMP *PROM* management.

PROMT *Predicasts overview of markets and terminology*. Databank, Predicasts Inc. (US).

PRONTO 1. programmable network telecommunications operating system. 2. keyboard trade name, not acronym.

PROP performance review for operating programs.

PRORA program for research on romance authors. Concordance software, University of Toronto (Canada).

PROSPEC PRO specification. Computerized archive management system, *PRO*.

PROSPO process systems program.

PROXI protection by reflection optics of xerographic images.

PRR pulse repetition rate.

PRRM pulse repetition rate modulation, data transmission.

PRS 1. partial response signalling. 2. Pattern Recognition Society (US). 3. pattern recognition system.

PRT program reference table.

PRTRC printer controller, computing.

PRU 1. packet radio unit. 2. physical record unit.

PS 1. parallel to serial. 2. problem specification. 3. processor status. 4. programming system.

PSA 1. parametric semiconductor amplifier. 2. psychological semantic analysis. Literary analysis software.

PSAM partitioned sequence access method.

PSB program specification block.

PSC 1. *Patent classification service, INPADOC*. 2. program sequence control.

PSCF processor storage control function.

PSCL programmed sequential control language.

PSD 1. packet switched data. 2. *Patent search documentation*. Databank, *EPO*. 3. post sending delay. 4. Printing Systems Division, Xerox Corporation. 5. program status double word.

PSDN public switched data network, telecommunications.

PSE packet switching exchange, telecommunications.

PSEC picoseconds (10^{-12} seconds).

PSI 1. Participation Systems Inc. Electronic communication company (US). 2. peripheral subsystem interface. 3. *Permuterm subject index, ISI*. 4. pressure-sensitive identification. 5. protosynthetic indexing. Program for indexing word occurrences.

PSIC process signal interface controller.

PSIEP project on scientific information exchange in psychology. 1960s *APA* project.

PSK 1. phase shift keying, telecommunications. 2. program selection key.

PSKM phase shift keyed modulation, telecommunications.

PSL 1. pocket select language, Burroughs (US). 2. problem specification language.

PSLI packet switch level interface.

PSM programming support monitor, *TI*.

PSN 1. packet switching node. 2. public switched network, telecommunications.

PSNS programmable sampling network switch.

PSP 1. packet switching processor. 2. presending pause, telecommunications. 3. programmable signal processor.

PSR 1. program status register. 2. program support representative.

PSS 1. packet switched system, telecommunications. 2. packet switchstream service, *BT*.

PSSC Public Service Satellite Consortium (US).

PSU 1. packet switching unit. 2. power supply unit. 3. program storage unit.

PSV pair shield video.

PSW 1. peripheral switching unit. 2. program status word.

PSYCINFO *Psychological information*. Database, *APA*.

PT 1. packet terminal, in *PSN*. 2. paper tape. 3. patentee. Searchable field, *ESA-IRS*. 4. phase type. Searchable field, Dialog *IRS*. 5. point. 6. printer terminal. 7. processing time. 8. punched tape.

PTA programmable translation array.

PTB physical transaction block.

PTC Pacific Telecommunications Council.

PTDOS processor technology disc operating system.

PTE page table entry.

PTF program temporary fix.

PTI program transfer interface.

PTL process and test language.

PTM 1. programmable terminal multiplexer, *TI*. 2. programmable timer module. 3. pulse time modulation.

PTM/OS programmable terminal monitor/operating system, *TI*.

PTOS paper tape oriented operating system.

PTP paper tape punch.

PTR paper tape reader.

PTS 1. phototypesetting. 2. *Predicasts terminal system*. Database on business (US).

PTSP paper tape software package, *DEC*.

PTT 1. post, telegraph and telephone authority. 2. post telephone or telex. Interlending service, Delft University of Technology (Netherlands). 3. push to talk.

PTTC paper tape and transmission code.

PTT/8 paper tape code on eight levels.

PTU package transfer unit.

PU 1. peripheral unit. 2. processing unit. 3. processor utility. 4. publisher. Searchable field, *BLAISE* and Dialog.

PUC 1. program under control. 2. Public Utilities Commission. Regulator of data traffic (US).

PUDOC Centrum voor Landbouw Publicaties Landbouw Documentatie. Centre for agricultural publishing and documentation. Host and information broker (Netherlands).

PUG *PASCAL* users' group.

PUL program update library.

PUMA programmable universal manipulator. Robot, Unimation Inc.

PUP peripheral universal processor.

PUT program update tape.

PV path verification.

PVC permanent virtual circuit. Permanently available connection in *PSN*.

PVD plan video display.

PVI programmable video interface.

PVP pipelined vector processor. Fast parallel processor.

PVR process variable record.

PVT page view terminal. Videotex terminal.

PW 1. *PRECIS* word. Searchable field, *BLAISE*. 2. private wire, telecommunications. 3. *Publishers' weekly*. Journal (US). 4. pulse width.

PWM pulse width modulation, equivalent to *PDM*.

PWS programmer work station.

PX private exchange, telecommunications.

PY publication year. Searchable field, *OLS*.

PZ performing organization zip code. Searchable field, Dialog *IRS*.

Q

QA 1. quality assurance. 2. topical qualifier abbreviation. Searchable field, *NLM*.

QADS quality assurance data system.

QAM 1. quadrature amplified modulation, data transmission. 2. queued access method. Data processing.

Q&D quick and dirty.

QAS question answering system.

QASP quality assurance surveillance plan.

Q-band radar frequency band.

QBANK *Kentucky quarterly state data bank*. Databank, *KEIS*.

QBE query by example. Search technique.

QBT quad bus transceiver.

QC 1. quality control. 2. queue control. 3. quick code. Searchable field, Dialog *IRS*.

QCC quality control centre.

QDPSK quaternary differential phase keying, telecommunications.

QE 1. qualifier entry version. Searchable field, Dialog *IRS*. 2. queue entry.

QED quick editor. Text editing software.

QF queue full.

QFM quantized frequency modulation.

QIO queue input/output.

QISAM queued indexed sequential access method, *IBM*.

QL 1. query language. 2. Quick Law Systems. Host to legal databases (Canada).

QLSA queue line sharing adaptor.

QN query normalization.

QNA *Quarterly national accounts*. Databank on *OECD* countries, *ADP*.

QOS quality of service.

QPAM quadrature phase and amplitude modulation.

QPR quadrature partial response.

QPRS quadrature channel modulation using *PRS*.

QPSK quaternary phase shift keying.

QRL quick relocate and link.

QRP query and reporting processor.

QRT queue run time.

QS 1. query system. 2. queue select.

QSA Quad synchronous adaptor, Perkin Elmer (US).

QSAM queued sequential access method, *IBM*.

QSL queue search limit.

QT qualifier type. Searchable field, *NLM*.

QTAM 1. queued telecommunications access method, *IBM*. 2. queued terminal access method.

Qtest quantitative test, of performance.

QTH queued transaction handling.

QUAD 1. quadrophonic, sound recording. 2. quadruplex. Video recording technique.

QUALTIS Quality Technical Information Service, *UKAEA*.

QUAM quantized amplitude modulation.

QUANTRAS question analysis transformation and search, University of Sheffield (UK).

QUEST 1. quality electrical system test. 2. query evaluation and search technique. 3. computer search language, not acronym.

QUESTEL Host to science, technology and business databases (France).

QUICK Quotation Information Center Kk. Originator and business databank (Japan).

QUICKTRAN quick *FORTRAN*. Computer programming language.

QUIC/LAW legal information service, Queen's University (Canada).

QUILL Queen's University interrogation of legal language. *IRS*, Queen's University of Belfast (UK).

QUIP 1. quick-inline package. *IC* packaging, Intel Corp. (US). 2. query interactive processor. 3. quota input processor.

QUIS Queen's University Information Systems, Queen's University of Belfast (UK).

QUOBIRD Queen's University online bibliographic information retrieval and dissemination, Queen's University of Belfast (UK).

QUODAMP *Queen's University databank on atomic and molecular physics*. Queen's University of Belfast (UK).

QVT Qume video terminal, Qume (UK) Ltd.

QWERTY keyboard layout, not an acronym.

R

R resistance.

RA 1. record address. 2. relocation address. 3. replacement algorithm. 4. return address. 5. ripple adder.

RAA 1. random access array. 2. remote access audio.

RAAR *RAM* address register.

RACE random access computer equipment.

RACF resource access control facility.

RACIC Remote Area Conflict Information Center, Battelle Memorial Institute (US).

RAD 1. random access device. 2. rapid access data drum, Xerox (US). 3. rapid access device. 4. rapid access disc.

RADA random access discrete address.

RADIALS *Research and development in information and library science*, now *Current research*, *LAPL*.

RADIR random access document indexing and retrieval.

RAG *ROM* address gate.

RAI random access and inquiry.

RAIN relational algebraic interpreter.

RAIR remote access immediate response.

RAK 1. read-access key. 2. remote access key.

RALF 1. rapid access to literature via fragmentation codes, E. Marck (FRG). 2. relocatable assembly language floating point.

RALU register arithmetic and logic unit.

RAM 1. random access memory. 2. reliability, availability and maintainability. Maintenance software, Burroughs (US).

RAMAC random access method of accounting and control.

RAMIO *RAM* plus input/output.

RAMIS rapid access management information system.

RAMP reliability and maintainability program.

RAMPI *Raw material price index*. Databank, also called *Pricedata*, Slamark International (Italy).

RANDAM random access non-destructive memory.

R&D research and development.

RAP 1. relational associative processor. 2. resident assembler program. 3. response analysis program.

RAPID random access personnel information disseminator.

RAPRA Rubber and Plastics Research Association. Originator and databank (US).

RAPS remote access power support.

RAR 1. return address register. 2. *ROM* address register.

RAROM *RAM* and *ROM*.

RAS 1. rapid access storage. 2. reliability, availability and serviceability.

RASI reliability, availability, serviceability and improvability.

RASIS reliability, availability, serviceability, integrity and security.

RASP 1. remote access switching and patching. 2. retrieval and statistics processing, *UKAEA*.

RASTAC random access storage and control.

RASTAD random access storage and display.

RATE remote automatic telemetry equipment.

RAW read after write.

RAX 1. remote access. 2. rural automatic exchange. Telecommunications.

RAYNET Raytheon Data Communications Network, Raytheon (US).

165

RB request block.

RBA relative byte address.

RBBS remote bulletin board system.

RBD reliability block diagram.

RBE remote batch entry.

RBF remote batch facility, *CDC*.

RBM 1. real-time batch monitor, Xerox (US). 2. remote batch module.

RBS remote batch system.

RBT remote batch terminal.

RBUPC *Research in British universities, polytechnics and colleges*. Database, *BLLD*.

RC 1. real circuit. 2. recipient city. Searchable field, Dialog *IRS*. 3. Regnecentralen computer. Computer series, Regnecentralen (Denmark). 4. remote channel. 5. remote computer. 6. remote concentrator. 7. remote control. 8. restrained cursor. 9. review classification. Searchable field, Dialog *IRS*.

RCA Radio Corporation of America.

RCB resource control block.

RC BASIC Regnecentralen *BASIC*.

RCC remote communications concentrator.

RC COMAL Regnecentralen *COMAL*.

RCF remote call forwarding.

RCI remote control interface.

RCIU remote computer interface unit.

RCP 1. receive clock pulse. 2. recognition and control processor.

RCS 1. reloadable control storage. 2. remote computing service. 3. remote control station.

RCU remote control unit.

RCV receive.

RCVR receiver.

RCW return control word.

RD 1. read data. 2. read direct. 3. receive data.

RDA run-time debugging unit.

RDAL representation dependent accessing language.

RDB relational data base.

RDC Regional Dissemination Center, *NASA*.

RDE remote data entry.

RDF 1. record definition field. 2. repeater distribution frame.

RDIU remote device interface unit.

RDL resistor diode logic.

RDOS 1. real-time disc operating system. 2. R disc operating system, Data General Nova Computers.

RDS Raytheon Data Systems (US).

RDT resource definition table.

RDTC remote distributed terminal controller.

RDY ready.

RDZ Ringier Dokumentationszentrum Zürich. Originator and database of newspaper articles and photographs (Switzerland).

REA Rural Electrification Administration. Constructs electric and telecommunications facilities (US).

READ 1. real-time electronic access and display. 2. relative element address designate. 3. remote electronic alphanumeric display.

REALCOM real-time communication.

REBI *Repertorio bibliografico Italiano*. Legal database, *CSC*.

REBIS *Repertorio bibliografico Straniero*. Legal database, *CSC*.

RECODEX *Report collection index*. Database of report literature, Studsvik Energitekniz (Sweden).

RECOL retrieval command language. Computer search language.

RECON 1. remote control online information service. 2. retrospective conversion, of non-*MARC* titles, *LC*.

REDAC real time data acquisition.

REDIC Red de Documentacion e Information de Colombia. Information and documentation network (Colombia).

REFLECS retrieval from the literature on electronics and computer science, *IEEE*.

REFLES reference librarian enhancement system. Online microcomputer system, University of California (US).

REFS remote entry flexible security.

REFSEARCH reference materials searching system, University of California at Berkeley (US).

REG *Règlement de la Chambre de communes*. Full text database of standing orders of Parliament (Canada).

REGIS 1. relational general information system. 2. remote graphics instruction set.

REIC *Rare earth information center*. Database, Ames Laboratory (US).

RELAY communications satellite, not an acronym (US).

RELCODE relative code, Sperry Rand (US).

RELI *Religion index*. Database, *ATLA*.

RELIPOSIS Research Liaison Panel on Scientific Information Services, British Gas.

REM 1. recognition memory. *ROM* used in *OCR*. 2. remark.

REMAP record extraction manipulation and print.

REMARC *Retrospective machine readable catalog*. Database, Carrolton Press Inc. (US).

REMDOS remote disc operating system, Datapoint (US).

REMSTAR remote electronic microfilm storage transmission and retrieval.

RENM request for next message.

REP re-entrant processor.

REPROM reprogrammable read-only memory.

REQ require.

RER residual error rate.

RES 1. remote entry service. 2. remote job entry system, *IBM*.

RESEDA 1. Réseau de Documentation en Économie Agricole. Originator and database on agricultural economics (France). 2. packet switched network (Spain).

RESORS *Remote scanning online retrieval system*. Database on telecommunications, Canada Centre for Remote Sensing.

RESP remote batch station program, Hitachi (Japan).

RESPONSA retrieval of specific portions from *Nuclear science abstracts*.

RETA retrieval of enriched textual abstracts. Information retrieval program.

RETMA Radio, Electronics, Television Manufacturers' Association, now *EIA* (US).

RETROSPEC retrospective search system, *INSPEC*.

REVS requirements engineering and validation system.

REWTEL Radio and Electronics World telecommunications. Microcomputer network, Radio and Electronics World publishers (UK).

REX real-time executive routine. Computing.

RF 1. radio frequency. 2. register file. 3. reporting file.

RFA remote file access.

RFD ready for data.

RFDU reconfiguration and fault detection unit.

RFG report format generator.

RFI 1. radio frequency interference. 2. request for information.

RFMS remote file management system.

RFNM ready for next message.

RFS random filing system.

RG release guard, telecommunications.

RGB red, green and blue.

RGP remote graphics processor.

RH report heading.

RHQ regional headquarters, British Telecom.

RHR receiver holding register.

RHT register holding time, telecommunications.

RHTM *Regional highway traffic model.* Databank for transport planning, Department of Transport hosted by *SIA* (UK).

RI 1. radio interference. 2. reliability index. 3. ring indicator, telecommunications.

RIC 1. *Rare earth information center.* Database, Iowa State University, Institute for Atomic Research and Ames Laboratory (US). 2. read-in counter.

RICASIP Research Information Center and Advisory Service on Information Processing (US).

RIE *Research in education.* Database, *ERIC*.

RIF reliability improvement factor.

RIGFET resistive insulated gate field effect transistor.

RIL representation independent language.

RILM *Repertoire international de littérature musicale.* Database on music, art and literature, *IAML* and *IMS*.

RIM 1. read-in mode. 2. resource interface module.

RIMS remote information management system.

RIN reference indicator number, *PRECIS*.

RINGDOC *Ring documentation.* Pharmaceuticals database, Derwent Publications (UK).

RIO 1. relocatable input/output. 2. remote input/output. 3. roll in only.

RIOT 1. *RAM* input/output timer. 2. retrieval of information by online terminal, *UKAEA*.

RIP 1. raster image processor. Assembles and stores images for visual display. 2. rest in proportion. Printing code.

RIPL representation independent programming language.

RIQS remote information query system. *IRS*.

RIR 1. relative index register. 2. *ROM* instruction register.

RIRA reports and information retrieval activity, Internal Revenue Service (US).

RIRO roll in/roll out. Storage allocation, computing.

RIS 1. raster input scanner, for pattern recognition. 2. remote information system.

RITA Recognition of Information Technology Achievement. Award (UK).

RITC Regional Information Technology Coordinators. Consultants appointed by government (UK).

RITL Royal Institute of Technology Library. Host and database originator (Sweden).

RJE remote job entry.

RJO remote job output.

RJP remote job processor.

RKNFSYS *Rock information system.* Databank, Carnegie Institution (US).

RL record length.

RLA remote line adaptor.

RLC 1. *ROM* location counter. 2. run length coding.

RLD relocation directory.

RLF reverse line feed. Control character.

RLG 1. release guard signal. 2. Research Libraries Group (US).

RLIN Research Libraries Information Network (US).

RLL relocating linking loader.

RLR record length register.

RLSD received line signal detector.

RM 1. register memory. 2. resource manager. 3. reliability and maintainability.

RMA 1. random multiple access. 2. reactive modulation amplifier.

RMB *ROM* memory band.

RMC Rod memory computer, *NCR*.

RMF resource measurement facility.

RML 1. relational machine language. 2. Research Machines Ltd. (UK).

RMM read mostly memory. *ROM* or *RAM* with safeguards against overwriting.

RMMU removable media memory unit.

RMON resident monitor, *DEC*.

RMOS refractory metal gate metal oxide semiconductor.

RMPI remote memory port interface.

RMS 1. record management system. 2. resource management support. 3. remote maintenance system. 4. resource management system. 5. root mean square, of transmission waves.

RMT remote.

RMTB reconfiguration maximum theoretical bandwidth.

RMW read modify write, computing.

RMX remote multiplexer.

RN 1. *CAS* registry number. Searchable field, Dialog and *NLM*. 2. reception node. 3. recipient name. 4. record number. Searchable fields, Dialog *IRS*. 5. report number. Searchable field, *OLS*.

RNAC remote network access controller.

RNC request next character.

RNP remote network processor.

RO 1. read only. 2. receive only. 3. register output.

ROAR Royal optimizing assembly routing, Royal McBee Corporation (US).

ROC remote operator's console.

ROF remote operator facility, Honeywell (US).

ROLS remote online system, Pertec (US).

ROM read-only memory. Computer memory.

ROMIO *ROM* plus input/output.

RONS read-only name store.

ROPES remote online print executive system.

ROS read-only storage.

ROSAR read-only storage address register.

ROSCOP *Report of observations samples collected by oceanographic programs.* Databank, *NODC*.

ROSDR read-only storage data register.

ROSE retrieval by online search, University of Pennsylvania (US).

ROTH read-only type handler.

ROTR read-only typing reperforator.

RP 1. read printer. 2. receive processor. 3. remote processor. 4. research problem area code. Searchable field, Dialog *IRS*.

RPC remote processor controller.

RPG report program generator.

RPI read punch and interpret.

RPL 1. remote program load. 2. reverse Polish logic, for arithmetic expression evaluation, computing.

RPM revolutions per minute.

RPN reverse Polish notation, for arithmetic expression evaluation, computing.

RPQ request price quotation.

RPR rings present. Searchable field, *SDC*.

RPS 1. real-time processing system. 2. records per sector. 3. remote printing system. 4. remote processing system. 5. revolutions per second. 6. rotational position sensing.

RPU regional processing unit.

RR 1. receive ready. 2. register to register. 3. return register.

RRA remote record address.

RRAR *ROM* return address register.

RRDS relative record data set.

RRE 1. receive reference equivalent, telecommunications. 2. Royal Radar Establishment. Research Centre (UK).

RRG resource request generator.

RRIS *Rail road research information service.* Database, *TRB*.

RROS resistive read-only storage.

RRP reader and reader-printer. For microform.

RRT relative retention time.

RS 1. real storage. 2. recipient state. Searchable field, Dialog *IRS*. 3. recommend standard. *EIA* prefix. 4. record separator. Control character. 5. register select. 6. register to storage. 7. remote site. 8. request to send.

RSA 1. remote session access. 2. requirements statement analyser.

RSC 1. remote store controller. 2. *Revised statutes of Canada.* Legal database, Department of Justice/*QL*.

RSCS remote spooling communications subsystem, *IBM*.

RSD 1. remote site data processing. 2. ring system descriptor. Searchable field, *SDC*.

RSF remote support facility.

RSP reader/sorter processor.

RSS 1. Raziskovalna Skupnost Slovenije. *INPADOC* operator (Yugoslavia). 2. route switching subsystem, telecommunications.

RSTS resource sharing time sharing.

RSU 1. register storage unit. 2. remote service unit.

RSVP remote system verification program, *HP*.

RSWB *Literaturinformation aus Raumordnung, Stadtebau, Wehnungswesen, Bauwesen.* Database on civil engineering and public administration, *IRB*.

RSX resource-sharing executive, *DEC*.

RT 1. real-time. 2. real-time operating system, *DEC*. 3. recipient type. 4. register transfer. 5. registration type. Searchable field, Dialog *IRS*. 6. related term. 7. remote terminal. 8. reperforator-transmitter. Teletypewriter.

RTA remote technical assistance, *CDC*.

RTAC real-time adaptive control.

RTAM remote terminal access method.

RTBM real-time bit mapping.

RTC 1. real-time clock. 2. remote terminal controller.

RTCA Radio Technical Commission for Aeronautics (Broadcasting).

RTCC real-time communications control.

RTCS 1. real-time communication system. 2. real-time composition system. Minicomputer-based system of Bedford Computer Corporation.

RTE remote terminal emulator.

RTECS *Registry of toxic effects of chemical substances.* Databank, *NIOSH*.

RTES real-time executive system, *SEMIS*.

RTIO 1. real-time input/output. 2. remote terminal input/output.

RTIP remote terminal interface package.

RTL 1. real-time language. 2. register transfer level. 3. resistor-transistor logic. 4. run-time library, Interdata (US).

RTLP reference transmission level point, telecommunications.

RTM 1. real-time module. 2. real-time monitor operating system, Systems Engineering Labs (US). 3. response time module.

RTMOS real-time multiprogramming operating system.

RTMS 1. real-time memory system. 2. real-time multiprogramming system.

RTN remote terminal network.

RTNR ringing tone no reply.

RTOP real-time optional processing.

RTOS real-time operating system, *CDC*.

RTP real-time processing.

RTR response time reporting.

RTS 1. rapid transmission and storage, Goldmark Corporation (US). 2. ready to send, data communications. 3. real-time subroutines. 4. real-time system. 5. remote testing system. 6. request to send. 7. Royal Television Society (UK).

RTTY radio teletypewriter.

RTU remote terminal unit.

R2E Réalisations et Études Électronique (France).

RTZ return to zero.

RU run unit.

RUF resource utilization factor.

RUIN Regional Urban Information Network, Washington DC (US).

RUM resource utilization monitor.

RUN rewind and unload.

RUNTR receive-only non-typing perforator.

RVA 1. recorded voice announcement, telecommunications. 2. relative virtual address.

RVI reverse interrupt. Control character.

RVT reliability verification tests.

R/W read/write.

RWED read/write extend delete.

RWM read/write memory.

RWR read/write register.

RWT right when tested.

RWTH Rheinische-Westfalische Hochschule. Database originator (FRG).

RX 1. receive. 2. receive mode. 3. receiver. 4. register to indexed storage.

RZ 1. Rechenzentrum. Computer centre (FRG). 2. return to zero.

RZ(NP) return to zero (non-polarized). Recording scheme.

RZ(P) return to zero (polarized). Recording scheme.

S

S 1. scalar. 2. second, of time. 3. section numbers. Searchable field, *ESA-IRS*. 4. Siemens. Unit of conductance. 5. storage.

SA 1. scholarship amount. Searchable field, Dialog *IRS*. 2. *Ship abstracts*. Database, Helsinki University of Technology, The Ship Research Institute of Norway and The Swedish Maritime Research Centre. 3. *Sociological abstracts*. Database, Sociological Abstracts Inc. (US). 4. source address. 5. sponsoring agency. Searchable field, Dialog *IRS*. 6. systems analysis. 7. systems analyst.

SAB stack access block.

SABE Society for Automation in Business Education (US).

SABIRC *Système automatique de bibliographie, d'information et de recherche en carcinologie*. Database and *IRS*, Institut Gustave-Roussy (France).

SABR symbolic assembler for binary relocatable (programs).

SABS Stanford automated bibliographic systems, Stanford University and *SSRC* (US).

SAC 1. semi-automatic coding. 2. serving area concept, in local line area (US). 3. special area code, telecommunications. 4. storage access channel. 5. store access controller, computing.

SACCS strategic air command control system (US).

SAD 1. store access director, *GEC* Computers (UK). 2. store address director.

SADF semi-automatic document feed.

SADP system architecture design package.

SADPO System Analysis and Data Processing Office, New York Public Library (US).

SADT structured analysis and design technique.

SAE abstracts *Society of Automotive Engineers abstracts*. Database, Society of Automotive Engineers (US).

SAF 1. segment address field. 2. short address form.

SAFAD Swedish Agency for Administrative Development. Coordinates and operates *LIBRIS*.

SAFF store and forward facsimile.

SAGACE *Système automatique pour la gestion et l'archivage des contes économiques*. Business and economics databank, *INSEE*.

SAGE semi-automatic ground environment. Air defence data system (US).

SAI sub-architectural interface.

SAIL Stanford artificial intelligence language.

SAINT 1. semi-automated indexing of natural language. 2. symbolic automatic integrator.

SAIT Southern Alberta Institute of Technology (Canada).

SAL 1. structured assembly language. 2. symbolic assembly language.

SALASSAH *Serials in Australian libraries: social sciences and humanities*. Database, *NLA*.

SALE simple algebraic language for engineers. Programming language.

SALINET Satellite Library Information Network. Library cooperative based on Denver Library School (US).

SAM 1. semantic analysing machine. 2. sequential access method, *IBM*. 3. serial access memory. 4. sort and merge. 5. subsequent address message, telecommunications. 6. systems adaptor module.

SAMANTHA system for the automated management of text in hierarchical arrangement.

SAMI Selling Areas Marketing Inc. Originator and databank (US).

SAMMIE system for aiding machine interaction evaluation.

SAMOS silicon and aluminium metal oxide semiconductor.

SAMP sense amplifier.

SAMS satellite auto-monitor system. Programming language.

SAMSON strategic automatic message switching operational network.

SAN standard address number, *BISAC* (US).

SANB *South African national bibliography*. Database, *CSIR*.

SANSS structure and nomenclature search system. Chemical information retrieval system.

SAOUG South African Online User Group.

SAP 1. structural analysis program. 2. symbolic address program.

SAPIR system of automatic processing and indexing of reports, University of California (US).

SAPLA Standing Advisory Panel on Library Automation, *LRCC*.

SAPRISTI système automatique de production d'information scientifique.

SAR 1. segment address register. 2. source address register. 3. storage address register.

SARIS South African Retrospective Information System.

SARM set asynchronous response mode.

SAROAD *Storage and retrieval of aerometric data*. Databank of pollutant levels, *EPA*.

SAS 1. selected applicant service, *IPG*. 2. switched access system.

SASI Shugart Associates system interface. Type of *SCSI*.

SASLIC Surrey and Sussex Libraries in Cooperation. Library cooperative (UK).

SAT system access technique, Sperry Univac (US).

SAT COM Committee on Scientific and Technical Communication. Former committee of *NAS* and *NAE* (US).

SATCOM satellite communication system, *RCA*.

SATF shortest access time first.

SATIS Scientific and technical information service, National Library of New Zealand.

SATO système d'analyse des textes par ordinateur. Analysis system for modern French text, Université du Québec à Montréal.

SATSTREAM satellite switchstream. Data transmission service, British Telecom.

SATT Strowger automatic toll ticketing, on Strowger telephone switching system (US).

SAU 1. smallest addressable unit. 2. system availability unit.

SAW surface acoustic wave. Utilized in microwave systems.

SAWIC South African Water Information Centre. Database originator.

SB 1. serial binary. 2. journal subset. Searchable field, *NLM*.

SBA shared batch area.

SBC 1. single board computer. 2. small business computer. 3. *Statutes of British Columbia*. Databank, Ministry of the Attorney General, British Columbia (Canada). 4. subtract contents. Assembler code. 5. system bus controller, computing.

SBCA sensor-based control adaptor.

SBCU sensor board control unit.

SBE system buffer element, computing.

SBI single byte interleaved.

SBIR storage bus in register.

SBR storage buffer register.

SBS Satellite Business Systems (US).

SBT 1. six-bit transcode. 2. surface barrier transistor.

SBT ASITO *Spravochnyy bank terminov avtomatizirannoy sistemy informatsionno-terminologicheskoyo obsluzhiveniya*. Multilingual terminology databank, *VNIIKI* (USSR).

SBU 1. station buffer unit, computing. 2. system billing unit.

SC 1. corporate search code. Searchable field, *ESA-IRS* and Dialog. 2. satellite computer. 3. section code. 4. section heading code. Searchable fields, Dialog *IRS*. 5. semiconductor. 6. sequence controller. 7. source code. 8. sponsor code. 9. subject category. 10. subject code. 11. system category code. Searchable fields, Dialog *IRS*.

SCA 1. short code address. 2. *Surface coating abstracts*. Database, Paint Research Association of Great Britain. 3. synchronous communications adaptor. 4. system control area.

SCADA supervisory control and data acquisition.

SCALD structural computer-aided logic design.

SCAM synchronous communications access method.

SCAN 1. Stock Market computer answering network (UK). 2. switched circuit automatic network.

SCANDOC Scandinavian Documentation Centre.

SCANNET Scandinavian network. Host and computer network (Sweden).

SCANP *Scandinavian periodicals index in economics and business*. Database, Helsinki School of Economics (Finland).

SCARS software configuration accounting and reporting system.

SCAT Schottky cell array technology, based on Schottky effect.

SCATS sequential controlled automatic transistor start.

SCATT scientific communications and technical transfer system, University of Pennsylvania (US).

SCAUL Standing Conference of African University Libraries.

SCB 1. segment control bit. 2. stack control block.

SCBS system control block.

SCC 1. satellite communication concentrator. 2. satellite communications controller. 3. secondary category code. Searchable field, *SDC*. 4. sectional classification code, *INSPEC*. 5. specialized common carrier, telecommunications. 6. sub-carrier channels, telecommunications. 7. synchronous communications controller.

SCCC single channel communications controller.

SCCDEST Steering Committee on Crossborder Data Exchange in Science and Technology (US).

SCCS source code control system.

SCCU single channel control unit.

SCE signal conversion equivalent.

SCEL Standing Committee on Education in Librarianship (UK).

SCEU selector channel emulator unit.

SCF 1. satellite control facility. 2. scientific computing feature, *ICL*.

SCFM sub-carrier frequency modulation.

SCI 1. *Science citation index*. Database, *ISI*. 2. system control interface.

SCIM selected categories in microfiche. Information dissemination service, *NTIS*.

SCIMP *Selective cooperative indexing of management periodicals*. Multilingual cooperative database, Helsinki School of Economics (Finland).

SCL 1. sequential control logic. 2. single channel monitoring. 3. Standard Chartered Leasing. Computer leasing company. 4. systems control language, *ICL*.

SCM 1. scientific calculator machine. Computer, *CIIHB*. 2. small core memory. 3. Society for Computer Medicine (US). 4. software configuration management.

SCN shortest connected network.

SCOBOL structured *COBOL*.

SCOCLIS Standing Conference of Cooperative Library and Information Services (UK).

SCOLCAP Scottish libraries cooperative automation project. Cooperative cataloguing project (UK).

SCONUL Standing Conference of National and University Libraries (UK).

SCOOP self-coupled optical pickup (Japan).

SCOPE 1. *Scholarly communication: online publishing and education.* Newsletter (US). 2. Standing Committee on Professional Education, *LA*. 3. supervisory control of program execution. 4. systematic computerized processing in cataloguing, Guelph University (Canada).

SCOPT Subcommittee on Programming Technology, Association of Computing Machinery (US).

SCORE system for computerized Olympic results and events, *TI* (US).

SCORPIO subject-content-oriented retriever for processing information online. Legal *IRS, LC*.

SCOUG Southern California Online User Group (US).

SCP 1. secondary cross-connection point, telecommunications. 2. supervisory control program, Burroughs (US). 3. symbol conversion program, *ITT*. 4. system control processor. 5. system control program. 6. systems control programming, *IBM*.

SCPC single channel per carrier. Telecommunications configuration.

SCPI *Small computer program index.* Periodical (UK).

SCR 1. scan control register. 2. silicon controlled rectifier. 3. single character recognition. 4. *Supreme Court reports.* Legal database, Department of Justice (Canada).

SCS 1. selected classification service, *IPG*. 2. separate channel signalling, telecommunications. 3. small computer system.

SCSI small computer system interface.

SCSU system control signal unit, telecommunications.

SCT 1. special characters table. 2. step control table.

SCTE Society of Cable Television Engineers.

SCU 1. station control unit. 2. storage control unit. 3. system control unit. 4. system/ memory control unit.

SCULL serial communication unit for long links.

SD 1. sales in millions of dollars. 2. search date. 3. send data. 4. serializer/deserializer. 5. start date. Searchable fields, Dialog *IRS*.

S/D signal to distortion ratio.

SDA 1. source data acquisition. 2. source data automation. 3. symbolic device address.

SDAID system debugging aids, *IBM*.

SDAL switched data access line.

SDB 1. segment descriptor block. 2. Societé de Banque de Données Bibliographiques. Originator of national bibliography (France). 3. storage data bus.

SDBP small database project.

SDBS Samson Database Services, *SDS*.

SDC 1. Scientific Documentation Centre Limited. 2. signal data converter. 3. Systems Development Corporation. Host (US).

SDCR source data communication retrieval.

SDD 1. selective dissemination of documentation, *AIAA*. 2. stored data description.

SDDL stored data definition language.

SDE source data entry.

SDF 1. standard data format. 2. supergroup distribution frame, of *FDM* system.

SDFS standard disc filing system.

SDI selective dissemination of information.

SDILINE *SDI* on *MEDLINE*.

SDIM *System for documentation and information in metallurgy*. Database, *BAM*.

SDIO serial digital input/output.

SDK system design kit.

SDL 1. software design language. 2. software development language, Burroughs (US). 3. system design language. 4. Systems Designers Ltd. (UK).

SDLC synchronous data link control.

SDM 1. semiconductor disc memory. 2. Staran debug module. 3. synchronous digital machine.

SDMA 1. shared direct memory access. 2. space division multiple access.

SDMS spatial data management system.

SDN synchronous digital network.

SDP source data processing.

SDR 1. signal to distortion ratio. 2. storage data recorder.

SDS 1. Samson Data Systemen. Host (Netherlands). 2. Scientific Data Systems Corporation (US). 3. simulating digital systems.

SDSI shared data set integrity.

SDSW sense device status word.

SDU 1. signal distribution unit. 2. source data utility. 3. station display unit.

SDW segment descriptor word.

SDX satellite data exchange.

SE 1. site engineer. 2. state. Searchable field, *SDC*. 3. systems engineer.

SEA 1. Societé d'Électronique et d'Automatisme (France). 2. static error analysis.

SEACOM South-east Asia Commonwealth cable, linking Singapore, Hong Kong and Australasia.

SEAL South-east Area Libraries. Library cooperative (UK).

SEAM software engineering and management.

SEARCH 1. system for electronic analysis and retrieval of criminal histories (US). 2. *Systemized excerpt abstracts and reviews of chemical headlines*. Database, International Business Data Inc. (US).

SEC 1. secondary electronic conduction. Video camera technology. 2. single error correcting.

SECAM séquential couleur à mémoire. Colour *TV* system.

SECDED single-bit error correction and double-bit error detection.

SECO sequential coding.

SECOURS système d'enseignement contrôlé par ordinateur et utilisant de resources satellites. Computer learning systems development programme, Université Laval (Canada).

SECU slave emulator control unit.

SEDIM single person general disc operating system, Intertechnique (France).

SEEA software error effects analysis.

SEED Strumech Engineering Electronic Developments. Hardware manufacturer.

SEF 1. software engineering facility. 2. standard external file. 3. storage extension frame.

SEG special effects generator. Video equipment for mixing images.

SEIRS *Suppliers and equipment information retrieval system*. Databank on aviation industry, *ICAO*.

SEITA Service de l'Exploitation Industrielle des Tabacs et des Allumettes. Database originator (France).

SEL Systems Engineering Laboratories. Computer manufacturer.

SELBUS Systems Engineering Laboratories data bus.

SELDOM selected dissemination of *MARC*, University of Saskatchewan (Canada).

SELECTAVISION selected television video disc system, *RCA*.

SEM standard electronic module.

SEMCOR semantic correlation. Computer-aided indexing system.

SEMI Semiconductor Equipment and Materials Institute.

SEMIS Société Européenne de Mini Informatique et des Systèmes, also *SEMS*.

SEMS see *SEMIS*.

SEN software error notification.

SENECA semantic networks for conceptual analysis. Database construction technique.

SENTOKYO Senmon Toshokan Kyogikai. Special Libraries Association (Japan).

SEP separation parameter.

SEQUEL structured English query language.

SER 1. sequential events recorder. 2. system environment recording.

SERC Science and Engineering Research Council (UK).

SERCNET SERC network. Computer network for *SERC* grantees.

SERF special extensive routine functions, *NCR*.

SERIX *Swedish environmental research index*. Database, Swedish Council of Environmental Information.

SERLINE *Serials online*. Database, *NLM*.

SES system external storage.

SESAM *Système électronique de sélection automatique des microfilms*. Database and *IRS* on welding, Centre de Recherches Scientifiques et Techniques de l'Industrie des Fabrications Métalliques (Belgium).

SESE single entry single exit.

SET stepped electrode transistor.

SETF Staran evaluation and training facility.

SETI Société Européenne pour le Traitement de l'Information.

SETM Société d'Études et de Travaux Mécanographiques.

SF 1. scale factor. 2. select frequency. 3. short format. 4. single frequency. 5. special features. Searchable field, Dialog *IRS*. 6. store and forward. 7. subfile. 8. substitute fragment. Searchable field, *SDC*.

SFA segment frequency algorithm.

SFC 1. sectored file channel. 2. sectored file controller. 3. selector file channel.

SFE smart front end.

SFF 1. San Francisco Federal Reserve Bank. Database originator (US). 2. standard file format.

SFF employment databank, *SFF*.

SFF financial databank, *SFF*.

SFF industrial databank, *SFF*.

SFL substrate fed logic.

SFM switching mode frequency multipliers.

SFP 1. security filter processor. 2. slack frame program.

SFS symbolic file support.

SFU special function unit.

SG system gain.

SGB 1. *Société Générale de Banque*. Databank on finance and economics (France). 2. Société Générale de Banque (de Belgique). Originator and databank on business and economics (Belgium).

SGB-DOC *Société Générale de Belgique – Documentation*. Database of documents related to Belgium, *CIGL*.

SGD self-generating dictionary.

SGJP satellite graphic job processor.

SGML standard generalized mark-up language. Proposed standard, *ANSI*.

SGU Sveriges Geologiska Undersokning. Geological survey of Sweden. Originates and operates *Grundva hendokumentation* databank on hydrogeology.

SH 1. *MESH* subheading. Searchable field, *NLM*. 2. sample and hold. 3. section heading. Searchable field, Dialog *IRS*. 4. session handler. 5. source handshake. 6. subject heading. Searchable field, *BLAISE*.

SHARES shared acquisition and retention system, New York Metropolitan Reference and Research Library Agency (US).

SHARP ships and analysis and retrieval project, US Navy.

SHAU subject heading authority unit. Software, Rand Corporation (US).

SHE *Subject headings for engineering*, Engineering Index Inc. (US).

SHF super high frequency (approx 10^{10} Hz).

SHFTR shift register.

SHIOER statistical historical input/output error rate utility, Sperry Univac (US).

SHIPDES *Ship descriptions*. Databank, Maritime Information Centre (Netherlands).

SHIRTDIF storage handling and retrieval of technical data in image formation.

SHL studio to head end link. Microwave link, *CATV* system.

SHOC software hardware operational control.

SI 1. serial input. 2. shift in. Control character. 3. storage immediate. 4. superintendent of document/item number. Searchable field, *SDC*. 5. système international. Measurement unit system.

SIA 1. Semiconductor Industry Association. 2. Service in Information and Analysis. Host (UK). 3. Standard Interface Adaptor.

SIAD Society of Industrial Artists and Designers (UK).

SIAM Society for Industrial and Applied Mathematics (US).

SIB serial interface board.

SIBIL système intégré pour les bibliothèques Universitaires de Lausanne. Interlibrary system (Switzerland).

SIC 1. Standard Industrial Classification. Indexing code for patent and business files. 2. *Système d'information conjonctorelle*. Databank on economic cycles, *INSEE*.

SICOB Salon International Informatique Télématique Communication Organisation du Bureau Bureautique. Business equipment exhibition (France).

SICOM Securities industry communications. Stock market information system, Western Union (US).

SICOMP Siemens computer. Computer series, Siemens (FRG).

SIC 72 *Standard industrial classification 72*. Databank on measures of business activity, *DOC*.

SID 1. signal identification, broadcasting. 2. Society for Information and Display (US). 3. sudden ionospheric disturbance, telecommunications. 4. Swift interface device, Swift Company. 5. syntax improving device.

SIDERAL *Système informatique de documentation en recherche aléatoire*. Essences et Lubricatifs Français – Enterprise de Recherches Pétrolières (France).

SIDES source input data edit system.

SIF storage interface facility.

SIFT software implemented fault tolerance.

SIG 1. special interest group, of eg *IIS, ASIS*. 2. sub-interface generator, *GEC* Computers (UK).

SIGACT Special Interest Group on Automata and Compatibility Theory, of *ACM*.

SIGGRAPH Special Interest Group on Graphics (US).

SIGI system for interactive guidance and information.

SIGLE system for information on grey literature in Europe. Information acquisition project.

SIGPLAN Special Interest Group on Programming Languages of *ACM*.

SIGSOC Special Interest Group on Social and Behaviour Science Computing.

SIIRS Smithsonian Institution information retrieval system (US).

SIL 1. scanner input language. 2. store interface link.

SILK system for integrated local communications. Network system, Hasler Company.

SILT stored information loss tree.

SIM synchronous interface module.

SIMD single instruction multiple data stream.

SIMILE simulation of immediate memory in learning experiments.

SIMON software interface monitor.

SIMP satellite information message protocol.

SIMPLE system for integrated maintenance and program language extension.

SIMS socio-economic information management system, University of Wisconsin (US).

SIN 1. subject indicator number, *PRECIS*. 2. symbolic integrator.

SINBAD *Système informatique pour banque de données*, Institut Nationale de la Santé et de la Recherche Médicale (France).

SINFDOC Statens Rad for Vetenskaplig Information och Dokumentation. State Council for Scientific Information and Documentation (Sweden).

SINTO Sheffield Interchange Organization. Library cooperative (UK).

SIO 1. serial input/output. 2. start input/output. 3. step input/output.

SIOC serial input/output channel.

SIP 1. scientific information processor, Honeywell (US). 2. single inline package. 3. software instrumentation package, Sperry Univac (US). 4. symbolic input program.

SIPROS simultaneous processing operating system, *CDC* Corporation (US).

SIPS 1. *State implementation plan system*. Databank of air pollution regulations, *EPA*. 2. statistical interactive programming system.

SIR 1. scientific information retrieval. Database management system (US). 2. segment identification register. 3. selective information retrieval. 4. semantic information retrieval, *MIT* (US). 5. service d'information rapide de l'Institut National de la Propriété Industrielle. Search service for users without terminals (France). 6. special information retrieval.

7. specification information retrieval system. 8. statistical information retrieval.

SIRA Institute Formerly, Scientific Instruments Research Association (UK).

SIRAID technical consultancy service, *SIRA Institute*.

SIRE Syracuse information retrieval experiments, Syracuse University (US).

SIRENE *Système informatisé pour la répertoire des enterprises et des établissements*. Databank on business statistics, *INSEE*.

SIRF *Statistiques et indicateurs des régions françaises*. Databank, *INSEE*.

SIRIO Communication satellite system (Italy).

SIRPI sistem automatizat de inmagazinare, regasire si presentare a informatulor. *IRS*, Centre for Information and Documentation for Agriculture and Silviculture (Romania).

SIS 1. scientific instruction set. 2. selected inventor service, *IPG*. 3. signalling interworking subsystem, telecommunications. 4. simulation interface system. 5. *Standards information system*. Databank, *NBS*. 6. supplier identification system, London Enterprise Agency and *UCSL*. 7. system interrupt supervisor.

SISDATA 1. single instruction single data stream. 2. *Statistical information system data*. Databank, Slamark International (Italy).

SIT 1. silicon intensified target, video camera technology. 2. special information tone, telecommunications. 3. static induction transistor.

SITA Société Internationale de Télécommunications Aéronautiques. International network for airline reservations.

SITC Salford Information Technology Centre. *IT* instructional centre for young people, Salford University (UK).

SITE satellite instructional television experiment, 1974 (India).

SITI Swiss Institute for Technical Information.

SITL static induction transistor logic. Integrated circuit system.

SITPRO Simplification of International Trade Procedures Board. Active in electronic trade data interchange standards.

SIU system interface unit.

SJF shortest job first.

SJP serialized job processor.

SKB skew buffer.

SKIL scanner keyed input language.

SKWOC structured keyword out of context. Indexing method.

SL 1. simulation language. 2. summary language. Searchable field, Dialog *IRS*. 3. system language.

SLA 1. shared line adaptor. 2. Special Libraries Association (US). 3. stored logic array. 4. synchronous line adaptor.

SLAM symbolic language adapted for microcomputers.

SLANG systems language.

SLC 1. selector. 2. simulated linguistic computer. 3. single line control. 4. systems life cycle.

SLCM software life cycle management.

SLCU synchronous line control unit.

SLD synchronous line driver.

SLDTSS single language dedicated time-sharing system.

SLE segment limits end.

SLI 1. suppress length indicator. 2. synchronous line interface.

SLIC 1. selected listing in combination. 2. subscriber loop interface circuit. Telecommunications integrated circuit.

SLICE 1. Surrey Library interactive circulation experiment, University of Surrey (UK). 2. system life-cycle estimation.

SLIH second level interrupt handler.

SLIP symmetric list processor.

SLIP 1. symbolic list processor. 2. symmetric list processor.

SLIPR source language input program.

SLISP symbolic list processing.

SLM synchronous line module.

SLO segment limits origin.

SLOCOP specific linear optimal control program.

SLOSRI shift left out/shift right in.

SLP segmented level programming.

SLR storage limits register.

SLS source library system.

SLSI super large-scale integration (100,000 or more components per chip).

SLT solid logic technology. Integrated circuit technology.

SLTF shortest latency time first.

SLU 1. serial line unit. 2. source library update. 3. special line unit, telecommunications.

SM 1. search month. Searchable field, Dialog *IRS*. 2. semiconductor memory. 3. set mode. 4. share of market. Searchable field, Dialog *IRS*. 5. structure memory. 6. synchronous modem.

S/M sort/message.

SMA surface mounting applicator. Machine for constructing *SMD*s.

SMAL 1. storage multiple access control. 2. store multiple access control. 3. structural macroassembly language.

SMART system for the mechanical analysis and retrieval of text.

SMB system monitor board.

SMC 1. storage module controller. 2. system monitor controller.

SMD 1. storage module drive. 2. surface-mounted drive.

SMDR station message detail recorder, Datapoint (US).

SMEM serial memory.

SMF 1. Société Mathématique de France. 2. systems management facilities, *IBM*.

SMI 1. simulated machine indexing. 2. static memory interface. 3. system memory interface.

SMIL Statistics and Market Intelligence Library, *DTI* (UK).

SMIP structure memory information processor.

SMIS symbolic matrix interpretation system.

SML symbolic machine language.

SMLCC synchronous multiline communications coupler.

SMM semiconductor memory module. 2. shared main memory. 3. shared multiport memory.

SMMC standard monthly maintenance charge.

SMP 1. symmetric multiprocessing. 2. system modification program.

SMPTE Society of Motion Picture and Television Engineers (US).

SMR series mode rejection.

SMRT single message rate timing.

SMS 1. Scientific Microsystems Inc. (US). 2. shared mass storage. 3. standard modular system. 4. storage management system.

SMT Société de Micro-informatique et de Télécommunications (France).

SMU 1. secondary multiplexing unit. 2. store monitor unit. 3. system maintenance unit. 4. system monitoring unit.

SMUT system for musical transcription, Indiana University (US).

SMX semi-micro xerography.

SN 1. *ISSN*. Searchable field, *ESA-IRS* and Dialog. 2. signal node. 3. source name. Searchable field, *SDC*. 4. *SSIE* number. 5. subject name. Searchable fields, Dialog *IRS*.

S/N signal-to-noise ratio. Telecommunications.

SNA systems network architecture. Computing.

SNAP 1. standard network access protocol. 2. system net activity program, Sperry Univac (US).

SNB *Statutes of New Brunswick*. Databank of Department of Justice, New Brunswick on *QL*.

SNBU switched network backup.

SNI selective notification of information.

SNICT Sistema Nacional de Informacao Cientifica e Tenologica (Brazil).

SNOBOL string oriented symbolic language.

SNP 1. statistical network processor. 2. synchronous network processor.

SNR signal-to-noise ratio. Telecommunications.

SNS 1. selected numeric service, *IPG*. 2. software notification service, *HP*.

SO 1. send only. 2. serial output. 3. shift out. 4. *Statutes of Ontario*. Full text database, *QL*. 5. source. Searchable field, Dialog and *SDC*.

SOA start of address.

SOAP symbolic optimum assembly programming, *IBM*.

SOB start of block.

SOCCER *SMART*'s own concordance constructor extremely rapid, Cornell University (US).

SOCCS study of computer cataloguing software (UK).

SOCIAL SCISEARCH *Social science citation index search*. Database on social sciences, *ISI*.

SOCO source code. Searchable field, *SDC*.

SOCRATES 1. system for organizing content to review and teach educational subjects, University of Illinois (US). 2. system for organizing current reports to aid technologists and scientists, Defense Science Information Service (Canada).

SOD 1. serial output data. 2. Superintendent of Documents, Government Printing Office (US). 3. system operational design.

SODIP Société pour la Diffusion de la Presse. Originator and database on serials (Belgium).

SOF start of frame.

SOGS *Saudi-oriented guide specifications.* Databank on construction specifications in Saudi Arabia, *DOC.*

SOH start of heading. Transmission control character.

SOL 1. *Solid waste.* Database, Environment Canada and *ORNL.* 2. source/object library.

SOLACE School of Librarianship Automatic Cataloguing Experiment, University of New South Wales (Australia).

SOLAR storage online automatic retrieval, Washington State University (US).

SOLINET South-eastern library network (US).

SOLINEWS *SOLINET newsletter.*

SOLO *SAMI online operations.* Database, *SAMI.*

SOLOMON simultaneous operation limited ordinal modular network.

SOLUG San Antonio On Line User Group.

SOM 1. small office microfilm systems. 2. standard online module, Fujitsu (Japan). 3. start of message.

SOMADA self-organizing multiple access discrete address.

SOP standard operating procedure.

SOR *Statutory orders and regulations.* Full text database, *QL.*

SORM set oriented retrieval module.

SOS 1. silicon on sapphire. Semiconductor technology. 2. sophisticated operating system, Apple III microcomputer.

SOS DOC Service d'orientation vers l'information et le documentation scientifique et technique. *EUSIREF* Centre (France).

SOT start of text.

SOTA state of the art.

SOT DAT *Source test data system.* Databank on air pollution, *EPA.*

SOUP software utility package.

SOURCE videotex system, not an acronym, Telecomputing Corporation of America.

SP 1. satellite processor. 2. send processor. 3. sequential processor. 4. serial to parallel. 5. space. Control character. 6. sponsor. 7. sponsoring program. Searchable fields, Dialog *IRS.* 8. stack pointer. 9. structured programming.

SPA Software Producers' Association (UK).

SPACE symbolic programming anyone can enjoy.

SPACECOMPS *Spacecraft components.* Databank, *ESA.*

SPADE single channel per carrier multiple access demand assignment equipment, telecommunications.

SPAM satellite processor access method.

SPAN span analysis. Text analysis using *COT* and *DAN.*

SPASM system performance and activity software monitor.

SPAU signal processing arithmetic unit.

SPB stored program buffer.

SPC 1. small peripheral controller. 2. Southern Pacific Communications (US). 3. stored program control.

SPCS 1. Standard and Poor's Compustat Services. Originator of business and economics databanks (US). 2. storage and processing control system.

SPD standard program device.

SPDM subprocessor with dynamic microprogramming.

SPE signal processing element.

SPEAKEASY high-level programming language, not an acronym.

SPEDE system for processing educational data electronically.

SPEED self-programmed electronic equation delineator.

SPF 1. standard program facility, *IBM*.
2. structured programming facility.

SPI 1. Semi Process Inc. Integrated circuit manufacturer (US). 2. shared peripheral interface. 3. single processor interface. 4. Société pour l'Informatique. Database originator (France).

SPIDAC specimen input to digital automatic computer.

SPIDEL Société pour l'Informatique et Documentation Électronique. Host (France).

SPIN 1. School Practices Information Network, *BRS*. 2. *Searchable physics information notes*. Database on physics, American Institute of Physics.

SPINDEX program package, *NHPC*.

SPIRES Stanford public information and retrieval system. Bibliographic database management system (US).

SPIS standard production information systems. Production management software, Trepel (UK).

SPK *Speaker's rulings*. Full text database of parliamentary Speaker's rulings, *QL* (Canada).

SPL 1. signal processing language.
2. simulation programming language.
3. source program library. 4. splice.
5. system program loader. 6. system programming language, *HP*. 7. systems programming language, Philips Company.

SPLC standard point location code.

S + DX speech plus derived telex. Simultaneous transmission of voice and low speed data.

SPM 1. scratch pad memory. 2. source program maintenance. 3. standard prototype microcomputer, Integrated Computer Systems (US). 4. subscriber's private meter, telecommunications.

SPMOL source program maintenance online.

SPN service protection network.

SPNB Security Pacific National Bank. Databank originator (US).

SPNS switched private network service.

SPO 1. separate partition option.
2. sponsoring organization. Searchable field, *SDC*.

SPOOF structure and parity observing output function.

SPOOL simultaneous peripheral operation online.

SPP 1. signal processing peripheral. 2. special purpose processor.

SPR storage protection register.

SPREAD support program for remote entry of alphanumeric displays, British Steel Corporation (UK).

SPRING switched private networking.

SPROM 1. slave programmable *ROM*.
2. switched programmable *ROM*.

SPS 1. serial parallel serial. 2. string process system. Word processing software.
3. symbolic program system. Computer system.

SPSS statistical package for the social sciences. Computer software for analysis of research results.

SPT 1. sectors per track. 2. structural programming technique.

SPTF shortest programming time first.

SPU slave processing unit.

SPUR 1. single precision unpacked rounded floating point package, Sperry Rand.
2. source program utility routine.

SPWM single-sided pulse width modulation, telecommunications.

SQA 1. software quality assurance.
2. system queue area.

SQAP Swedish Question Answering Project, Research Institute of National Defence (Sweden).

SQD signal quality detector.

SQL structured query language, formerly *SEQUEL*.

SQL/DS *SQL*/data system, *IBM*.

SQNCR sequencer microprogram.

SR 1. send receive. 2. shift register. 3. sorter reader. 4. special register. 5. speech recognition. 6. status register. 7. storage register. 8. switch register.

SRAM 1. semi *RAM*. 2. static *RAM*.

SRB sorter reader buffered.

SRC 1. *Les statutes revisés du Canada*. Legal database, Department of Justice/*QL*. 2. Science Research Council, now *SERC* (UK). 3. synchronous remote control.

SRCNET *SRC* network. Now *SERCNET*.

SRCR system run control record.

SRE 1. sending reference equivalent, telecommunications. 2. single region execution.

SRF 1. software recording facility. 2. software recovery facility. 3. sorter reader flow.

SRI Stanford Research Institute. Databank originator and *IT* consultant (US).

SRL 1. Science Reference Library (UK). 2. shift register label. 3. structural return loss, telecommunications.

SRM 1. short range modem. 2. system resource manager, *IBM*.

SRP shared resources programming.

SRS 1. selective record service. Cataloguing service. 2. slave register set.

SRZ Satz Rechen Zentrum. Host (FRG).

SS 1. select standby. 2. start/stop. 3. storage to storage. 4. system supervisor.

SSA single line synchronous adaptor.

SSAC signalling system alternating current, telecommunications.

SSB single sideband, transmission.

SSBAM single sideband amplitude modulation.

SSBM single sideband modulation.

SSBSC single sideband suppressed carrier.

SSB-SC/AM single sideband suppressed carrier amplitude modulation.

SSC 1. *SCANNET* Service Centre. 2. sector switching centre, telecommunications. 3. signalling and supervisory control. 4. Solid State Scientific Inc. Microprocessor manufacturer (US). 5. station selection code. 6. system support controller.

SSCI *Social science citation index*. Database, *ISI*.

SSDA synchronous serial data adaptor.

SSDC signalling system-direct current, telecommunications.

SSDD single-sided double density, magnetic disc.

SSDR steady state determining routine.

SSF symmetrical switching function.

SSG symbolic stream generator.

SSI 1. small scale integration. 2. Special Secretariat for Informatics (Brazil). 3. synchronous systems interface.

SSIE *Smithsonian science information exchange*. Databank of research in progress, Smithsonian Institution (US).

SSL 1. scientific subroutine library. 2. Software Sciences Limited (UK). 3. software specification language. 4. storage structure language. 5. system specification language.

SSLC synchronous single-line controller.

SSM 1. semiconductor storage model. 2. single sideband signal multiplier, telecommunications. 3. small semiconductor memory.

SSMF signalling system multi-frequency, telecommunications.

SSN 1. segment stack number. 2. switched service network, telecommunications.

SSP 1. signalling and switching processor. 2. Society for Scholarly Publishing (US). 3. system status panel. 4. system support processor. 5. system support program.

SSR 1. solid state relay. 2. system status report.

SSRC Social Science Research Council (UK).

SSRC survey archive Databank, *SSRC*.

185

SSRS source storage and retrieval system.

SSP Structured Sound Synthesis Project. Computer music synthesis project, *SSRC* of Canada and Computer Systems Research Group, University of Toronto (Canada).

SST 1. Software Sciences Teleordering. Largely owned by Thorn-EMI, now Teleordering Ltd (UK). 2. subscriber transferred signal, telecommunications. 3. synchronous system trap. 4. system segment table.

SSTDMA satellite switched time division multiple access. Satellite communications technique.

SSTF shortest seek time first.

SSTV slow scan television.

SSU 1. subscriber's switching unit, telecommunications. 2. subsequent signal unit. Group of bits in *CCS*.

ST 1. source and time frame. Searchable field, *NLM*. 2. stereochemical descriptor. Searchable field, Dialog *IRS*. 3. supplementary terms. Searchable field, *SDC*. 4. system table, computing.

STA 1. store answer. Assembler code. 2. store in memory from accumulator. Assembler code.

STAC software timing and control.

STAIRS storage and information retrieval system. *IR* software, *IBM*.

STAIRS/VS *STAIRS*/virtual storage.

STAM shared tape allocation manager.

STAR 1. *Science and technology aerospace reports, NASA*. 2. *Serials titles automated records*. Agricultural database, *IBM*. 3. self-testing and repairing. 4. string array processor, *CDC*. 5. system to automate records. *IRS*, Cuadra Associates (US).

STARFIRE systems to accumulate and retrieve financial information with random extraction.

STAT citations to statutes at large. Searchable field, *SDC*.

STAT MUX statistical multiplexer, of *TDM* system.

STB segment tag bits.

STC 1. Society for Technical Communication (US). 2. Standard Telephone and Cable. *ITT* affiliate (UK). 3. standard transmission code. 4. Storage Technology Corporation (US).

STCB sub task control block.

STD 1. Society of Typographic Designers (UK). 2. subscriber trunk dialling. Telecommunications.

STDM 1. statistical time division multiplexing. 2. synchronous time division multiplexing.

STDN Spacecraft Tracking and Data Network.

STE supergroup translating equipment, in *FDM* system.

STELLA Satellite Transmission Experiment Linking Laboratories.

STEP standard tape executive system, *NCR*.

STF 1. short title file. Bibliographic record. 2. stereochemistry fragment. Searchable field, *SDC*. 3. supervisory time frame.

STI scientific and technical information.

STIBOKA Stichting voor Bodemkartering. Soil survey institute. Originator and databank on soils (Netherlands).

STINFO 1. Scientific and Technical Information Officers (group) (US). 2. *STI* program. 1960s project, US Department of Defense.

STIR statistics indexing and retrieval project, Loughborough University of Technology (UK).

STIS Specialized textile information service. Database and *IRS*, Shirley Institute (UK).

STISI Service du Traitement de l'Information et des Statistiques Industrielles. Databank originator and operator (France).

STL 1. Schottky transistor logic, based on Schottky effect. 2. studio transmitter link, telecommunications.

STLB Schottky *TTL* bipolar. *LSI* technology.

STM 1. scientific technical and medical. 2. short-term memory. 3. STM Publishers. International organization (Netherlands).

STN switched telecommunications network.

STO 1. segment table origin. 2. *Standing orders*. Full text database of orders of Parliament, *QL* (Canada).

STOCKLIBRIS information service for industry, Stockport Public Library (UK).

STOIAC Static Technology Office Information Analysis Center, Battelle Memorial Institute (US).

STOL systems test and operation language.

STOQ storage queue.

STORES syntactic tracer organized retrospective enquiry system, Instituut voor Wiskunde, Informatiewerking en Statistiek (Netherlands).

STORET *Storage and retrieval for water quality data*. Databank, *EPA*.

STP system test procedure.

STPL standard test processing language. Proposed international standard.

STR 1. segment table register. 2. side-tone reduction, telecommunications. 3. software trouble reporting service, Data General. 4. speed-tolerant recording. 5. strobe. 6. symbol time recovery. 7. synchronous transmit receive.

STRAIN structural analytical interpreter.

STRC Science and Technology Research Center, University of North Carolina (US).

STRN standard technical report number, *NTIS*.

STRUDL structural design language.

STS space time space. Digital switching structure.

STSC Scientific Time Sharing Corporation. Host (US).

STSI scientific technical and societal information.

STT 1. seek time per track. 2. single transmission time.

STTC Schottky transistor-transistor logic, based on Schottky effect.

STU 1. Styrelsen for Teknisk Utvekling. National Board for Technical Development. Originator and database (Sweden). 2. system time unit. 3. system transmission unit.

STW Communications Satellites (China). Not an acronym.

STX start of text character. Control character.

SU 1. selectable unit. 2. signalling unit. 3. single user. 4. storage unit.

SUB 1. subroutine, within program. 2. substitute. Control character.

SUBTIL synthesized user-based terminology index language, McGill University (Canada).

SUI standard universal identifier.

SULIRS Syracuse University libraries' information retrieval system (US).

SULIS Sulzer Literaturverteilung und Sortierung. Literature sorting and dissemination system (Switzerland).

SUM systm utilization monitor.

SUMEXAIM Stanford University medical experiment – applications of artificial intelligence to medical research (US).

SUMMIT Sperry Univac minicomputer management of interactive terminals.

SUN 1. *SPINDEX* Users' Network. 2. symbols, units, nomenclature.

SUPARS Syracuse University *Psychological abstracts* retrieval service (US).

SURGE sorting, updating, report generating etc. Compiler, *IBM*.

SUS speech understanding system.

SUSE *Système unifié de statistiques d'enterprises*. Databank on company finance, *INSEE*.

SUSY Such Systems. Press agency (FRG).

SV 1. single value. 2. status valid.

SVA shared virtual area.

SVC 1. supervisor call instruction. 2. switched virtual call, telecommunications. 3. switched virtual circuit.

187

SVD 1. Schweizerische Vereinigung für Dokumentation. Documentation Association (Switzerland). 2. simultaneous voice and data.

SVDF segmented virtual display file.

SVI service interception.

SVM silicon video memory.

SVP service processor.

SVR super video recorder, Grundig.

SVS single virtual storage, *IBM*.

SVSF Sveriges Vetenskapliga Specialbiblioteks Forening. Special Research Libraries Association (Sweden).

SVT system validation testing.

SW 1. software. 2. subject word. Searchable field, *BLAISE*. 3. switch.

SWA system work area.

SWADS scheduler work area data set, *IBM*.

SWALCAP South West Academic Libraries Cooperative Automation Project (UK).

SWAMI software-aided multifont input. OCR software.

SWAP standard wafer array programming.

SWB *Summary of world broadcasts*. Database, *BBC* via Datasolve.

SWE status word enable.

SWEDIS *Swedish drug information system*. Databank originated and operated by Swedish National Board of Health and Welfare.

SWI software interrupt.

SWIFT 1. selected words in full title. 2. significant word in full title. 3. Society for Worldwide Interbank Financial Transactions. International banking network.

SWIFTLASS signal word index of field and title literature abstracts specialized search.

SWIFTSIR signal word index of field and title science information retrieval.

SWIRS *Solid waste information retrieval system*. Database, *EPA*.

SWOV Stichting Wetenschappelijk Onderzoek Verkeersveiligheid. Institute for road safety research. Operator of *IRRD* database (Netherlands).

SWP Stiftung Wissenschaft und Politik. Database originator and operator (FRG).

SWRA *Selected water research abstracts*. Database, *WRSIC*.

SWRI Southwest Research Institute. Database originator (US).

SWUCNET Southwest universities computer network (UK).

SWULSCP Southwest university libraries systems cooperative project. Now *SWALCAP*.

SX simplex. Single direction transmission.

SXS step by step switch.

SY 1. search year. 2. synonyms. Searchable fields, Dialog *IRS*.

SYCLOPS *SYFA* current logic operating system.

SYFA system for access, Computer Automation Inc. (US).

SYLCU synchronous line control unit.

SYLK symbolic link format.

SYMBIOSIS system for medical and biological information searching, New York State University (US).

SYN synchronous idle. Transmission control character.

SYNC synchronizing pulses. Telecommunications.

SYSGEN 1. system generation. 2. system generator program.

SYSIN system input.

SYSOUT system output.

SYSTRAN translation system. Machine translation system: English, French, Russian, Spanish.

SYU synchronization signal unit.

SZ sponsoring organization zip code. Searchable field, Dialog *IRS*.

T

T tera (10^{12}).

TA 1. Telecom Australia. Common carrier. 2. terminal address. 3. (journal) title abbreviation. Searchable field, *NLM*. 4. title annotation. Searchable field, *SDC*. 5. traduction automatique. Automatic translation. 6. Triumph Adler computer series.

TAB tabulation.

TABS *Tailored abstracts, INSPEC.*

TABSIM tabulating simulator.

TABSOL tabular systems oriented language, *GEC*.

TAC 1. *TELENET* access controller. 2. terminal access controller. 3. translator assembler compiler.

TACOS tool for automatic conversion of operational software.

TACRAHD tactical routing indicator look-up and header preparation device, *CDC*.

TACS total access communication system.

TACT terminal activated channel test.

TAD 1. terminal address designator. 2. transaction application drive, Computer Technology Inc. (US).

TAF transaction facility, *CDC*.

TAG time automated grid.

TAL terminal application language.

TALINET telefax library network. Experimental network (US).

TALON Texas Arkansas Louisiana Oklahoma New Mexico. Regional medical library programme, *NLM*.

TAM 1. telecommunications access method. 2. terminal access method.

TAMOS 1. terminal automatic monitoring system. 2. terminal auto-operator and monitor system, Nixdorf (FRG).

TAP 1. terminal access processor. 2. test assistance program, Sperry Univac (US). 3. *Thermodynamics and physical properties package*. Databank, University of Houston (US).

TAR 1. *Tax advice rulings*. Full text database, *QL* (Canada). 2. terminal address register. 3. track address register.

TARDIS time and attendance recording analysis, *BLSL*.

TARGET Machine-aided translation system for English, French, German, Spanish. Not an acronym.

TARP test and repair processor.

TAS test and set.

TASC tabular sequence control.

TASCON television automatic sequence control.

TASI time assignment speech interpolation. Telephone transmission timesharing technique.

TASO Television Allocation Study Organization (US).

TASS teleprinter automatic switching system.

TAT transatlantic telephone cable.

TAU 1. thesaurus alphabetical up to date, *NASA*. 2. trunk access unit.

TB tone burst.

TBC 1. time base corrector. Early digital processing device. 2. time base corrector. Element in video recording system.

TBEM terminal-based electronic mail.

TBM tone burst modulation.

TBR table base register.

TC 1. tag code. Searchable field, Dialog *IRS*. 2. tape command. 3. taxonomic code. Searchable field, *SDC*. 4. term coordination. 5. terminal concentrator. 6. terminal

191

control. 7. time clock. 8. treatment code. Searchable field, *ESA-IRS*. 9. trunk control.

TCA 1. Telecomputing Corporation of America. Operator of The Source videotex system. 2. terminal communications adaptor.

TCAM telecommunications access method, *IBM*.

TCAM-IMS/VS *TCAM* – information management system/virtual storage, *IBM*.

T-carrier system telecommunications carrier system. Long distance *PCM* system (US).

TCB 1. task control block. 2. transfer control block.

TCBH time consistent busy hour, telecommunications.

TCC 1. Technical Change Centre (UK). 2. toll centre code, telecommunications. 3. transmission control characters.

TCIS telex computer inquiry service.

TCL 1. terminal command language, Applied Digital Data Systems (US). 2. terminal control language.

TCM 1. telecommunications monitor, Siemens (FRG). 2. terminal to computer multiplexer. 3. thermal conduction module.

TCMF touch calling multi-frequency, telecommunications.

TCP 1. task control program. 2. terminal control program. 3. transmission control protocol. 4. transmitter control pulse.

TCP/IP transmission control protocol/ internet protocol, *ARPANET*.

TCR tape cassette recorder.

TCS 1. telecommunications control system, Toshiba (Japan). 2. telecommunications system, Siemens (FRG). 3. telex communications service, Western Union (US). 4. terminal communications subsystem. 5. terminal control system, *HP*. 6. transaction control system, Hitachi (Japan).

TCTS Trans-Canada Telephone System. Telecommunications network.

TCU 1. tape control unit. 2. terminal control unit. 3. timing control unit. 4. transmission control unit.

TD 1. termination date. Searchable field, Dialog *IRS*. 2. text and data exchange. 3. top down. 4. transmission distributor. 5. transmit data. 6. transmitter distributor.

TDA tunnel diode amplifier.

TDB terminology database.

TDCK Technisch Documentatiecentrum voor de Krijgsmacht. Technical Documentation Centre for the Army (Netherlands).

TDCS time division circuit switching, telecommunications.

TDE total data entry.

TDF 1. Télédiffusion de France. Broadcasting agency, including teletext. 2. transborder data flow. 3. trunk distribution frame, telecommunications.

TDG test data generator.

TDI telecommunications data interface.

TDL transaction definition language.

TDM 1. template descriptor memory. 2. time division multiplexing. Transmission technique.

TDMA 1. tape direct memory address. 2. time division multiple access, telecommunications.

TDOS tape *DOS*.

TDPL top down parsing language.

TDR 1. tape data register. 2. time delay receiver. 3. tone dial receiver. 4. transactional document recorder, Bell and Howell (US).

TDRS 1. text data retrieval system. 2. trucking and data relay satellite.

TDS 1. transaction distribution system. 2. transaction driven system.

TDTL tunnel diode tunnel logic.

TDX time division exchange.

TE transverse electric. Wave propagation mode.

TEAM 1. teleterminals expandable added memory. 2. terminology evaluation and acquisition method. Machine-aided translation system.

TEBOL terminal business-oriented language.

TEC triple erasure connection.

TECHNONET Industrial technical information network, National Research Council (Canada).

TED 1. television disc. Videodisc system, *TelDec*. 2. *Tenders electronic daily*. Database of public works tenders in European Community.

TEDS twin exchangeable disc storage.

TEG Training and Education Group, *LA*, formerly *LEG*.

TEGAS test generation and simulation.

TEL task execution language.

TelDec Telefunken-Decca (FRG-UK).

TELDOK telefunken dokumentationssystem. Search software.

TELECOM 1. telecommunications. 2. database, *SDBS*.

Teledata viewdata system (Norway).

TELEDOC *Telecommunications documentation*. Database, *CNET*.

TELEFAX facsimile transmission services (France and FRG).

TELEFAX 201 facsimile transmission service (Netherlands).

TELEMAIL electronic mail service (US).

TELENET telecommunication network (US).

TELEPACK telecommunications service (US).

TELESET viewdata system (Finland).

TELETEL brand name of *ANTIOPE* videotex system (France).

TELEX teletype exchange. Teleprinter service, Western Union.

TELIDON viewdata system (Canada).

TELTIPS technical effort locator and technical interest profile system, US Army.

TEMA Telecommunication Engineering and Manufacturing Association (UK).

TEMPOS timed environment multipartitioned operating system.

TEP 1. terminal error program. 2. test executive processor.

TEPOS test program operating system.

TERMDOK *Terminology databank* for *MAT*, *TNC*.

TERMINOQ *Banque de terminologie de Québec*. Multilingual terminology databank, Office de Langue Français, Gouvernement de Québec (Canada).

TERMIUM *Banque de terminologie de l'Université de Montréal*. Multilingual terminology databank (Canada).

TES 1. text editing system. 2. time encoded speech, telecommunications.

TESLA Technical Standards for Library Automation Committee, *LITA*.

TEST *Thesaurus of engineering and science terms*, Engineers' Joint Council (US).

TEX 1. tau epsilon chi. Computer composition system, Knuth (US). 2. teleprinter exchange, Western Union (US).

TEXPLOT Texas Instruments plotter. Computer graphics equipment, *TI* (US).

TEXTLINE *Text online*. Business information database, Finsbury Data Services Ltd. (UK).

TF tape feed.

TFM tape file management.

TFMS text and file management system.

TFR transaction formatting routines.

TFS tape file supervisor.

TFT thin film technology, microelectronics.

TFVA Training Film and Video Association (UK).

TG 1. Telecom Gold. *BT* electronic mail system. 2. terminator group.

TGI *Target group index*. Databank on market research information, Axiom Market Research Bureau Inc. (US).

TH transmission header.

THC Thomson CSF (France).

THD total harmonic distortion.

THESEE *Thesaurus pour l'électricité et l'électronique.*

THIS Tobacco and Health Information Services, University of Kentucky (US).

THOR tape handling optional routines, Honeywell.

THP terminal handling processor.

THQ telecommunications headquarters.

THR transmitter holding register.

3C Catalog Card Corporation of America.

3M Minnesota Mining and Manufacturing Co. Database originator (US).

TI 1. Texas Instruments (US). 2. terminal interface. 3. title. Searchable field, *ESA-IRS* and Dialog. 4. title words. Searchable field, *SDC*.

TIB Technische Informationsbibliothek. Library and document delivery centre (FRG).

TIC 1. Technical Information Center, *DOE* (US). 2. terminal identification code.

TICCIT time shared interactive computer controlled information television, Mitre Corporation and Brigham Young University (US).

TICS telecommunication information control system.

TIDF trunk intermediate distribution frame, telecommunications.

TIDMA tape interface direct memory access.

TIE 1. terminal interface equipment. 2. time interval error, telecommunications.

TIF 1. tape inventory file. 2. terminal independent format.

TIGER total information gathering and reporting, *ICL* (UK).

TIGS terminal independent graphics system.

TIH trunk interface handler.

TIL Technical Indexes Ltd. (UK).

TIM 1. table input to memory. 2. total information management. *DBMS* for micros.

TIMS transmission impairment measuring set.

TIM/TOM table input to memory/table output from memory.

TINDX *TI* indexed access method.

TINET transparent intelligent network.

TINTO Tieteellisen Informoinnin Neuvosto. Council for Scientific and Technical Information (Finland).

TIO 1. terminal input/output. 2. test input/output.

TIOC terminal input/output controller.

TIOT task input/output table.

TIOWQ terminal input/output wait queue.

TIP 1. technical information program. 2. telefiche image processor, for facsimile transmission. 3. telenet interface processor. 4. terminal interface processor. 5. terminal interface program. 6. transaction interface package, Sperry Univac (US). 7. transaction interface processor.

TIPS 1. text information processing system. 2. thousands of instructions per second. Unit of computer processing speed.

TIP TOP tape input/tape output.

TIQ task input queue.

TIR 1. target instruction register. 2. total internal reflection in fibre optic cables.

TIRC Toxicology Information and Response Center, US Atomic Energy Commission and *NLM*.

TIRP textile information retrieval program, *MIT* (US).

TIS *The Information System*. Database system.

TIT test item taker.

TITUS *Traitement de l'information textile universelle et sélective*. Database and *IRS*, Institut Textile de France.

TIU terminal interface unit.

TJID terminal job identification.

TKK (Helsingen) Teknillisen Korkeakoulun Kirjasto. Helsinki University of Technology Library, information broker and database originator and operator (Finland).

TKO trunk offering, telecommunications.

TL transaction language.

TLB transaction look-aside buffer.

TLC task level controller.

TLCT total life-cycle time.

TLMS tape library management system.

TLN trunk line network.

TLP telephone line patch.

TLS 1. tape librarian system. 2. Tekniska Litteratursallskapet. Society for Technical Documentation (Sweden). 3. Total Library System, *OCLC*.

TLSA transparent line sharing adaptor.

TLU table look-up.

TM test mode.

TMA Telecommunications Management Association. Division of Institute of Administrative Management (UK).

TMC tape management catalogue.

TMCS Toshiba minicomputer complex system.

TMF transmission monitoring facility.

TMIC Toxic Material Information Center, US Atomic Energy Commission and National Science Foundation.

TMP terminal monitor program.

TMR triple modular redundancy.

TMS 1. tape management system. 2. telex management systems, Racal Information Systems (UK). 3. transmission measuring set.

TMU 1. test maintenance unit. 2. transmission message unit.

TMXO tactical miniature crystal oscillator.

TN 1. task number. Searchable field, *SDC*. 2. terminal node. 3. trade name. Searchable field, Dialog *IRS*.

TNC 1. Tekniska Nomenklaturcentralen. Centre for technical terminology. Operates *TERMDOK* terminology bank (Sweden). 2. terminal network controller. 3. transport network controller.

TNDC Thai National Documentation Centre.

TNF third normal form, in database structure development.

TNO Organisatie voor Toegepast-Natuurwetenschappelijk Onderzoek. Organization for applied scientific research. Database originator and information broker (Netherlands).

TNTID Tsentar za Naucho-Teknicheska Informatsii i Dokumentatsiia. Centre for Scientific and Technical Information and Documentation (Bulgaria).

TNYT *The New York Times online*. Electronic publication, *NYTIS* (US).

TO *Telegraph offices*. Databank, *ITU*.

TOADS terminal oriented administrative data system.

TOCS terminal oriented computer system.

TOD time of day.

TODS test oriented disc system.

TOL test oriented language.

TOLAR terminal online availability reporting.

TOLTE teleprocessing online test, Fujitsu (Japan).

TOLTEP teleprocessing online test executive program, *IBM* (US).

TOLTS total online testing system, Honeywell (US).

TOOL test oriented operator language.

T_1, T_2 etc. *PCM* line arrangement series.

TOPCAT Texas onboard program of computer assisted training. For *CAT* of mariners, *UWIST* and Texaco Overseas Tankship Co. (US).

TOPIC Information system, not an acronym, London Stock Exchange (UK).

TOPS 1. telephone order processing system. 2. teletype optical projection system. Radar plotting system. 3. testing and operating system. 4. time-sharing operating system. 5. total operations processing system.

TOS 1. tape operating system, *IBM*. 2. temporarily out of service. 3. top of stack pointer, computing.

TOSCIN top of stack control word.

TOSL terminal oriented service language.

TOSS terminal oriented support system.

TOXBACK *Toxicology information backup*. Offline service of *TOXLINE*.

TOXBIB *Toxicity bibliography*. Database, *NLM*.

TOXICON *Toxicology information conversation online network*. Now *TOXLINE*.

TOXLINE *Toxicology information online*. Database, *NLM*.

TOXTIPS *Toxicology testing in progress*. Databank, *NLM*.

TP 1. terminal processor. 2. test point. 3. text processing. 4. text processor. 5. transaction processing.

TPA transient program area.

TPAM teleprocessing access method.

TPD thermoplastic photoconductor device.

TPFI terminal pin fault insertion.

TPG telecommunication program generator.

TPI tracks per inch, of magnetic storage devices.

TPL 1. table producing language. 2. terminal processing language. 3. test processing language.

TPMM teleprocessing multiplexer module.

TPMS transaction processing management system.

TPNS teleprocessing network simulator.

TPRC Thermophysical Properties Research Center (US).

TPS 1. tape programming system, *IBM*. 2. telecommunications programming system. 3. terminal polling system, *TI*. 4. transaction processing system.

TPU 1. tape preparation unit. 2. task processing unit. 3. telecommunications processing unit.

TQE time queue element.

TR 1. tape resident. 2. *Technical report program*. Database, *DDC*. 3. *Tekniska rapporter*. Database of reports held by AB Atomenergie (Switzerland). 4. translation register.

TRACS transport and road abstracting and cataloguing system, Transport and Road Research Laboratory (UK).

TRAMP timeshared relational associative memory program.

TRAN transmit.

TRANSDOC *Transport documentation*. Database, *OECD*.

TRANSINOVE *Transfer of innovation*. Database, France.

TRANSMUX transmission multiplexer.

TRANSPACK Telecommunications network (US).

TRAWL tape read and write library.

TRB Transport Research Board. Database originator (Canada).

TRC 1. Technology Reports Centre. Division of *DOI*. Broker and database operator (UK). 2. transmit receive control. 3. transverse redundancy check.

TRC-AS transmit/receive control unit – asynchronous start/stop.

TRC-SC transmit/receive control unit – synchronous character.

TRC-SF transmit/receive control unit – synchronous framing.

TRE transmit reference equivalent, telecommunications.

TREM tape reader emulator module.

TRF tuned radio frequency.

TRIAL technique for retrieving information from abstracts of literature.

TRIB transfer rate of information bits.

TRIM tailored retrieval and information management.

TRIPS 1. *TALON* reporting and information processing system, *NLM* (US). 2. travel information processing system. Travel agent videodisc project (US).

TRIS *Transport research information services.* Databank, US Department of Transportation.

TRISNET *TRIS* network.

TRK track.

TRL transistor resistor logic.

TRM terminal response monitor.

TROPAG *Tropical agriculture*. Database, Royal Tropical Institute (Netherlands).

TROPO tropospheric scatter communications. Long distance microwave transmission method.

TROS transformer read only storage.

TRQ task ready queue.

TRR tape read register.

TRRL Transport and Road Research Laboratory. Database operator (UK).

TRS 1. Tandy Radio Shack. 2. telephone repeater station. 3. transpose.

TRT turn round time.

TRTL transistor-resistor transistor logic.

TS time sharing.

TSA time slot access.

TSAM time series analysis and modelling. *ADP* software for *IR*.

TSB terminal status block.

TSC 1. Technical Support Center, *IBM* (US). 2. three state control. 3. time-sharing control. 4. totally self-checking. 5. transit switching centre, telecommunications. 6. transmitter start code, teletypewriter systems.

TSCA *Toxic Substances Control Act*. Database of all chemicals in TSCA initial inventory of 1976 and first supplement, *NHLI*.

TSCT time-sharing control task, Fujitsu (Japan).

TSE terminal source editor.

TSI 1. test structure input. 2. transmitting subscriber's identification.

TSID track sector identification.

TSIU telephone system interface unit.

TSL test source library.

TSM terminal support module.

TSN task sequence number.

TSO 1. telephone series observations. 2. time-sharing option.

TSODB time series oriented database.

TSOS time-sharing operating system.

TSP total systems performance.

TSPS traffic service position system. Telecommunications.

TSR temporary storage register.

TSS 1. terminal security system. 2. terminal send side. 3. terminal support subsystems, *SEL*. 4. time-sharing system. 5. time-shared system. 6. toll switching system.

TSSMCP time-sharing system message control program, Fujitsu (Japan).

TST time-space time. Digital switching structure. 2. transaction step task.

TSU trunk switching unit, telecommunications.

TSW 1. task status word. 2. telesoftware. Software distributed via videotex. 3. the searchers workbench. Prototype *IR* enhancement machine, University of Illinois (US).

TSX time-sharing executive, Modular Computer Systems.

TT 1. teletypewriter. 2. title terms. Searchable field, Pergamon Infoline and *SDC*. 3. top term, in thesaurus hierarchy. 4. transaction terminal.

T/TAL Tandem Computers transaction application language, Tandem Computers (US).

TTD temporary text delay.

TTDF *Tariff and trade data files*. Databank, *GATT*.

TTD keyterm index *Textile technology digest keyterm index*. Databank, *ITT*.

TTDL terminal transparent delay language.

TTHM Turk Teknik Haberlesme Merkezi. Technical Information Centre (Turkey).

TTL 1. title. Searchable field, Pergamon Infoline. 2. transistor-transistor logic.

TTLS *TTL* Schottky.

TT-N *Tidningarnas Telegrambyra nyhetsbanken*. Full text news database, Tidningarnas Telegrambyra (Sweden).

TTR 1. transmission test rack. 2. trunk test rack, telecommunications.

TTS 1. teletypesetter. 2. teletypesetting. 3. teletypesetting code. 4. transaction terminal system. Banking terminal system, Fortronic.

TTSPN two-terminal series parallel networks.

T²L transistor-transistor logic, also *TTL*.

TTY teletypewriter.

TTYC *TTY* controller.

TTYPP teletype point-to-point online communications driver, *NCR*.

TU 1. tape unit. 2. timing unit.

TUA Telecommunications Users Association (UK).

TUCC Triangle Universities Computation Center, of North Carolina, North Carolina State and Duke Universities (US).

TUF transmitter underflow.

TULSA Database on petroleum, University of Tulsa (US).

TURDOK Turk Bilismensel ve Teknik Dokumentasyon Merkezi. National Science and Technology Documentation Centre (Turkey).

TV 1. television. 2. threshold value. 3. transfer vector.

TVA Tennessee Valley Authority. Databank originator (US).

TVC control tag vector.

TVDR tag vector display register.

TVR tag vector response.

TVRO television receive only. Receiving site for communication satellite signal.

TVS transient voltage suppressor.

TV-SAT/TFD Communications satellite project (France and FRG).

TVT 1. television terminal. 2. television typewriter.

TVW tag vector word.

TW text word(s). Searchable field, *BLAISE* and *NLM*,

TWA 1. transaction work area. 2. two-way alternate.

TWB typewriter buffer.

TWP twisted wire pair.

TWR tape write register.

TSW two-way simultaneous.

TWT travelling wave tube.

TWTA *TWT* amplifier.

TWX Teletypewriter Exchange Service (US and Canada).

TX 1. taxonomic descriptor. Searchable field, Dialog *IRS*. 2. telephone exchange. 3. telex. 4. (with suffix) *TI* executive operating system. 5. transmit. 6. transmit mode. 7. transmitter.

TXD telephone exchange – digital.

TXDS *TI* diskette operating system.

TXE telephone exchange – electronic.

TXK telephone exchange – crossbar.

TXS telephone exchange – Strowger.

TYME-GRAM Tymnet telegram. Electronic mail service, Tymnet Inc. (US).

U

UA user area. Information storage location.

UAC uninterrupted automatic control.

UACN unified automated communication network.

UAIDE Users of Automatic Information Display Equipment (US).

UAP 1. universal availability of publications. 2. user area profile.

UAR unit address register.

UART universal asynchronous receiver transmitter.

UAX unit automatic exchange, telecommunications.

UBA 1. Umwelt Bundesamt. Database originator (FRG). 2. unblocking acknowledgement signal, telecommunications.

UBC 1. universal bibliographical control. 2. universal book code. Notional standard book code.

UBL unblocking signal, telecommunications.

UC 1. unichannel. 2. universal decimal code. Searchable field, Dialog *IRS*. 3. up converter.

UCB 1. unit control block. 2. universal character buffer.

UCC 1. unified classification code, *INSPEC*. 2. United Computer Corporation (US). 3. Universal Copyright Convention. 4. University Computing Co. International computer bureau and host.

UCF utility control facility.

UCI user class identifier, for data security.

UCLA University of California at Los Angeles. Databank originator.

UCM universal communications monitor.

UCMP *Union catalog of medical periodicals*. Database, Medical Library Center of New York (US).

UCP uninterruptable computer power.

UCS 1. universal call sequence. 2. universal character set. 3. universal communications subsystem.

UCSD-p University of California at San Diego-p. Operating system.

UCSL Unilever Computer Services Ltd.

UCSTR universal code synchronous transmitter receiver.

UCW 1. unit control word. 2. Union of Communication Workers, formerly *UPW* (UK).

UD update. Searchable field, *NLM*.

UDAC user digital analog controller.

UDAS unified direct access standard, *ICL*.

UDC 1. universal decimal classification. Library classification. 2. universal digital control.

UDE universal data entry.

UDL uniform data language.

UDR universal document reader.

UDS 1. Uniscope display system, Sperry Univac (US). 2. universal distributed system.

UDTS Universal Data Transfer Service (Philippines).

UE user equipment.

UEIS United Engineering Information Service (US).

UERPIC *Underground Excavation and Rock Properties Information Center*. Database, *CINDAS*.

UET universal emulating terminal.

UF utility file.

UFAM universal file access method.

UFC universal frequency counter.

201

UFD user file directory.

UFEDS unified fixed exchangeable disc coupler.

UFI 1. upstream failure indication, telecommunications. 2. usage frequency indicator. 3. user-friendly interface.

UFM user to file manager.

UFOD Union Française des Organismes de Documentation. Association of Documentation Organizations (France).

U format unknown format.

UFP utility facilities program.

UHF ultra high frequency (approx 10^9 Hz).

UHL user header label.

UHR ultra high reduction, greater than 90x, of microfilm.

UHRF ultra high resolution facsimile.

UIC user identification code.

UIG user instruction group.

UIO universal input/output.

UIOC universal input/output controller.

UIR user instruction register.

UJCL universal job control language.

UKAEA UK Atomic Energy Authority. Databank originator and operator.

UKASE University of Kansas automated serials system (US).

UKC *United Kingdom current MARC file*. Database, *BLAISE*.

UKCIS UK Chemical Information Service. Information broker, databank originator and host.

UKCTRAIN UK catalogue training. *UKMARC* training file.

UKITO UK Information Technology Organization.

UKLDS UK library database system. Proposed national system for catalogue records.

UKMARC *UK MARC*. Database of UK publications, *BLAISE*.

UKOLUG United Kingdom Online Users Group.

UKR *UK retrospective*. Database, *BLAISE*.

UL 1. uncontrolled (index) language. 2. upper list. 3. user language.

ULA uncommitted logic array.

ULAP Universitywide library automation program, now *DLA*, University of California.

ULC 1. uniform loop clock. 2. universal logic circuit. 3. upper and lower case, printing.

ULISYS universal library system. Circulation package, Universal Library Systems.

ULM 1. universal line multiplexer. 2. universal logic module.

ULSCS University of London shared cataloguing system.

ULSI ultra large scale integration, of integrated circuits.

ULTRA LSI *ULSI*.

U-matic Video cassette, Sony (Japan).

UMC unibus micro channel.

UMF ultra micro fiche.

UMI 1. *UMF*. 2. University Microfilms International. Microfilm and database producer (US).

UMIST University of Manchester Institute for Science and Technology. Databank originator and research institute (UK).

UMLC universal multiline controller.

UMPLIS Umweltplanungsinformations System. Environmental planning information system (FRG).

UMS universal memory system.

UMS/VS universal multiprogramming system/virtual storage, Mitsubishi (Japan).

UMTRIS *Urban mass transportation research information service*. Database, US Department of Transportation.

UN United Nations.

UNALC user network access link control.

UNCHS *UN* Centre for Human Settlements. Information broker (Kenya).

UNCM user network control machine.

UNCOL universal computer oriented language.

UNDEX *UN* documents index.

UNDIS *United Nations documentation information system.* Database on law and public administration, Dag Hammarskjold Library, *UN* (US).

UNEP United Nations Environmental Programme. Database originator (Kenya).

UNESCO United Nations Education, Scientific and Cultural Organization. Database originator and operator (France).

UNIBUS universal bus, *DEC*.

UNICAT *National union catalogue of monographs* (South Africa).

UNICOMP universal compiler.

UNIMARC universal *MARC*.

UNIQUE uniform inquiry update element.

UNISIST United Nations information system in science and technology. Information programme, *UNESCO*.

UNISTAR user network for information storage, transfer, acquisition and retrieval.

UNIVAC universal automatic computer. Early computer, Remington Rand Corporation, later computer series, Sperry Rand Corporation (US).

UNIVERSE Universities expanded ring and satellite experiment. Cooperative communication project (UK).

UNV Utviklingsselskapet for Naeringsliv pa Vestlandet. Database originator and operator (Norway).

UP 1. uniprocessor. 2. update code. Searchable field, *ESA-IRS, SDC* and Pergamon Infoline.

UPACS universal performance assessment and control system.

UPC 1. unit of processing capacity. 2. universal peripheral controller. 3. universal product code. Bar code.

UPC-E universal product code – Europe.

UPI 1. United Press International. Newswire service, produces full text database (US). 2. universal personal identifier.

UPL user programming language, Burroughs (US).

UPP universal *PROM* programmer.

UPS 1. uninterruptable power supply. 2. universal processing system.

UPSI user program sense indicator.

UPT user process table.

UPU Universal Postal Union. Postal cooperation organization, *UN*.

UQT user queue table.

UR 1. unit record. 2. unit register.

URA user requirements analysis.

URC unit record control.

URL user requirements language.

URP unit record processor.

US unit separator. Control character.

USAM unique sequential access method.

USART universal synchronous/asynchronous receiver/transmitter.

USASCII US *ASCII*, more commonly *ASCII*.

USASCSOCR USA standard *OCR* character set.

USASCU United States of America standard code for information interchange. Computer code.

USAI USA Standards Institute.

USB upper side band.

USBE Universal Serials and Book Exchange. Library supplier (US).

USCA *United States contract awards.* Databank, Washington Representative Services (US).

USCLASS *US classifications.* Database on patents, Derwent Publications.

USDA United States Department of Agriculture. Database originator.

203

USDA/CRIS *USDA current research information system.* Databank, *USDA*.

USEMA US Electronic Mail Association.

USER user system evaluator.

USGCA *US Government contract awards.* Now *USCA*.

USGS US Geological Survey. Databank originator.

USI 1. universal software interface. 2. user system interface.

USICA US International Communications Agency.

USIO unlimited sequential input/output.

USIS *Utviklingsselskapets sekundaer-informasjons system.* Databank of information contacts, *UNV*.

USITA United States Independent Telephone Association. Organization of telephone companies.

USNCFID US National Committee of *FID*.

USPA *US patents.* Database, Derwent Publications.

USPO US Post Office.

USPS US Postal Service.

USPSD *US Political science documents.* Database, University of Pittsburgh (US).

USP70 *US patents 70.* Database, Derwent Publications.

USP77 *US patents 77.* Database, Derwent Publications.

USR user service routine, *DEC*.

USRPF (1/2/3) *US request for proposals.* Databank of US Federal Government requests for R&D proposals, Washington Representative Services.

USRT universal synchronous receiver/transmitter.

UT 1. uncontrolled terms, for online searching. 2. up time. Operational time. 3. user terminal.

UTA user transfer address.

UTC Utilities Telecommunications Council (US).

UTD universal transfer device.

UTICS University of Texas Institute for Computer Science (US).

UTLAS University of Toronto library automation systems. Library cooperative and network (Canada).

UTOL universal translator oriented language.

UTRIN Ústvedi Technického Rozvoja a Informaci. *COMPENDEX* operator (Czechoslovakia).

UTS 1. universal terminal system, Sperry Univac. 2. universal timesharing system, Amdahl Company.

UTS/VS universal timesharing system/virtual storage, Mitsubishi (Japan).

UTTC universal tape to tape converter.

UUT unit under test.

UV ultraviolet. Range of electromagnetic spectrum

UVLI Ustav Vedeckych Lekarskych Informaci. Institute for medical information. Database operator (Czechoslovakia).

UVPROM ultra violet *PROM*.

UVTEI Ústvedi Vedeckych Technikych a Ekonomickych Informaci. Centre for Scientific, Technical and Economic Information (Czechoslovakia).

UWA user working area.

UWIST University of Wales Institute of Science and Technology (UK).

V

V 1. volt. 2. prefix to data communications standards, *CCITT*.

VA 1. virtual address. 2. volt ampere.

VAA voice access arrangement.

VAAP Vsesoyuznoe Agentstro po Avtorskim Pravam. Copyright agency (USSR).

VAB voice answer back.

VAC 1. value-added carrier. 2. voltage-alternating current.

VACC value-added common carrier.

VADAC voice analyser data converter.

VADC video analog to digital converter.

VAM virtual access method.

VAMP vector arithmetic multi-processor.

VANS value-added network services.

VA-NYTT *Vatten-och avloppsnytt*. Database on the environment, K-Konsultt (Sweden).

VAS 1. value-added service. 2. vector addition system. 3. videodisc authoring system, *WICAT*.

VAT 1. virtual address translator. 2. voice activation technology.

VAU vertical arithmetic unit.

VB voice band, telecommunications.

VBI vertical base organization and maintenance processor.

VBOMF virtual base organization and maintenance processor.

VBP virtual block processor.

VBS Vision Business Systems Ltd (UK).

VC 1. verification condition. 2. virtual circuit.

VCA 1. valve control amplifier. 2. Viewdata Corporation of America. 3. voice connecting arrangement. 4. voltage controlled amplifier.

VCBA variable control block area.

VCC video compact cassette. Video recorder, Philips.

VCCS Video and Cable Communications Section of the *ALA*.

VCF voltage controlled filter.

VCG verification condition generator.

VCO voltage controlled oscillator.

VCP virtual control panel.

VCR video cassette recorder.

VCRO validity check and readout.

VCS 1. validation control system. 2. video communications system. 3. video computer system. Computer supporting video display.

VCTCA virtual channel to channel adaptor.

VD virtual data.

VDA verbal delay announcement, telecommunications.

VDAM virtual data access method.

VDB 1. vector data buffer. 2. video display board.

VDC 1. video data controller. 2. voltage – direct current.

VDE voice data entry.

VDE/h Verein Deutscher Eisenhuttenleute. Databank originator (FRG).

VDETS voice data entry terminal system.

VDG 1. video data generator. 2. video display generator.

VDI 1. video display input. 2. video display interface. 3. visual display input.

VDP video data processor.

VDR voice digitization rate.

VDS voice data switch.

VDT 1. video data terminal. 2. video display terminal. 3. visual display terminal.

VDU 1. video display unit. 2. visual display unit.

VDUC *VDU* controller.

VEA virtual effective address.

VENUS valuable and efficient network utility service.

VET visual editing terminal.

VETDOC *Veterinary documentation.* Database, Derwent Publications (UK).

VF 1. voice frequency (300-3,000 Hz). 2. voltage to frequency.

VFB vertical format buffer.

VFC 1. variable file channel. 2. vertical format control. 3. voltage to frequency converter, used in analog to digital conversion.

VFCT voice frequency carrier telegraph.

VFL variable field length.

VFMED variable format message entry device.

VFO 1. variable frequency oscillator. 2. voice frequency oscillator.

V-format variable length file format.

VFT voice frequency telegraph(y).

VFU 1. vertical format unit. 2. vocabulary file utility.

VGAM vector graphics access method.

VGU video generation unit. Intelligent *CRT* controller.

VHA very high accuracy.

VHD video high density. Video disc system, *JVC*.

VHF very high frequency (approx 10^8 Hz).

VHLL very high level language.

VHM virtual hardware monitor.

VHOL very high order language.

VHPIC very high performance integrated circuits.

VHR very high reduction, of microfilm (61x to 90x).

VHS video home system. Video cassette system, *JVC*.

VHSI very high speed integration.

VHSIC very high speed integrated circuits.

VIA Videotex Industry Association Ltd (UK).

VIBL variable intensity back lighting.

VIC virtual interaction controller.

VICAM virtual integrated communication access method, Sperry Univac (US).

VICAR video image communication and retrieval.

VIDEO *VORTEX* interactive data entry operation, Sperry Univac (US).

VIDIAC video input to automatic computer.

Viditel viewdata system (Netherlands).

Vidon viewdata system (Canada).

VIMTPG virtual interactive machine test program generator, Sperry Univac (US).

VINE *Very informal newsletter.* Library automation newsletter, University of Southampton (UK).

VINITI Vsesojoznyi Institut Naucno-techniceskoi Informacii. All Union Institute for Scientific and Technical Information (USSR).

VIO 1. video input/output. 2. virtual input/output.

VIP 1. variable information processing. Information retrieval system, Naval Ordnance Laboratory (US). 2. vector instruction processor. 3. videodisc innovation project, Utah State University (US). 4. visual information processor. Software, Digital Research. 5. visual information projection, Honeywell (US).

VIPS variable item processing system.

VIR vertical interval reference. Lines in *TV* transmission reserved for teletext.

VIROC visible system of information retrieval by optical coordination.

VISC video disc.

VITAL *VAST* interface test application language.

VIU video interface unit. For computer control of videodisc players.

VIURAM *VIU RAM*.

VLB *Verzeichnis lieferbarer Bücher*. Books in print (FRG).

VLF 1. variable length field. 2. very low frequency (approx 10^4 Hz).

VLP video long player. Video disc system, Philips/*MCA*.

VLS virtual linkage system.

VLSI very large scale integration. Integrated circuits.

VLSW virtual line switch.

VM 1. virtual machine. 2. virtual memory.

VMA 1. valid memory address. 2. virtual machine assist, *IBM*. 3. virtual memory allocation.

VMAPS virtual memory array processing system.

VM/BSE virtual machine/basic system extension, *IBM*.

VMCB virtual machine control block.

VMCF virtual machine communication facility.

VME virtual machine environment, *ICL*.

VMID virtual machine identifier.

VML Virtual Microsystems Ltd.

VMM virtual machine monitor.

V-mode records records of variable length.

VMOS 1. vertical metal oxide semiconductor. 2. virtual memory operating system, Sperry Univac (US). 3. V-type metal oxide semiconductor.

VM/SE virtual machine/system extension, *IBM*.

VM/SP virtual machine/system product, *IBM*.

VMT 1. variable microcycle timing. 2. video matrix terminal.

VNC voice numerical control.

VNIIKI Vsesoyuznii Nauchno-Issledovatel'skii Institut Tekhnicheskoi Informatsii, Klassifikatskii i Kodirovaniya. All Union Research Institute for Engineering Information, Classification and Coding (USSR).

VNIIMI Vsesoyuznii Nauchno-Issledovatel'skii Institut Meditsinskoi i Medikotekhnicheskoi Informatsii. All Union Scientific Research Institute for Medical and Medico-technical Information (USSR).

VNL via net loss. Signal loss over telecommunications trunks.

VNLF via net loss factor.

VNN vacant national number. Signal, international telecommunications.

VOCODER voice coder. Telecommunications.

VOGAD voice operated gain adjustment device. Telecommunications.

VOL volume.

VOM volt ohm meter.

VORTEX versatile omnitask real-time executive, Sperry Univac (US).

VOS virtual operating system.

VOSC *VAST* operating system code.

VOTERM voice terminal. Voice recognition terminal.

VOTES database on voting records, US Congress.

VP 1. vector processor. 2. virtual control program, National CSS (US). 3. virtual processor.

VPAM virtual partitioned access method.

VP and VLE *(Organic systems) vapour pressures and vapour liquid equilibria*. Databank, *NPL*.

VP/CSS virtual program/conversation software system, National CSS (US).

VPN virtual page number.

VPS virtual programming system, National CSS (US).

VPSW virtual program status word.

VPZ virtual processing zero.

VR validation and recovery.

VRAM 1. variable rate adaptive multiplexing. 2. video *RAM*.

VRC 1. vertical redundancy check. 2. visible record computer.

VRF vertical random format. Graphics format, Versatec Company.

VRM visible record machine.

VROM video *ROM*.

VRP visual record printer.

VRU voice response unit.

VRX virtual resource executive, *NCR*.

VRX-MP *VRX* – multiprocessor.

VS 1. vestigial sideband. 2. virtual storage. 3. virtual system. 4. vocal synthesis.

VSAM 1. virtual sequential access method. 2. virtual storage access method. 3. virtual system access method.

VSB vestigial sideband modulation.

VSBS very small business systems.

VSC virtual subscriber computer.

VSDM variable scope delta modulation.

VSE virtual storage extended, *IBM*.

VSE/AF virtual storage exhibit/advanced function, *IBM*.

VSI virtual storage interrupt.

VSL variable specification list.

VSMF visual search on microfilm.

VSN volume serial number.

VSPC virtual storage personal computing, *IBM*.

VSS 1. video storage system. 2. virtual storage system, *SEMIS*. 3. voltage to substrate and sources, microelectronics.

VS-SC vestigial sideband suppressed carrier.

VSWR voltage standing wave ratio.

VT 1. vertical tabulate. 2. vertical tabulation.

VTAC video timing and control.

VTAM 1. virtual telecommunications access method, *IBM*. 2. virtual terminal access method.

VTB *Verfahrenstechnische Berichte*. Database on chemical engineering, Farbenfabriken Bayer AG (FRG).

VTDI variable theshold digital input.

VTI 1. Statens Vag-och Trafikinstitut. National Road and Traffic Research Institute. Database originator and operator (Sweden). 2. video terminal interface.

VTLS Virginia Technical Library System, Virginia Polytechnic Institute (US).

VTOC volume table of contents.

VTP 1. viewdata terminal program. 2. virtual terminal protocol.

VTR 1. video tape recorder. 2. video tape recording.

VTS Viewscan text system. Aid for partially sighted, *WISA* Company.

VTT Technical Research Centre of Finland. *IT* research centre, information broker and database originator.

VTTC video tape time-code. Location coding on video tape.

V2000 video cassette recorder system, Philips.

VU 1. voice unit. 2. volume unit. Measures of signal amplitude.

VUBIS Vrije Universiteit Brussels and Interactive Systems Incorporated.

VWS variable word size.

W

W 1. watt. Unit of power. 2. write.

WAA *World aluminium abstracts*. Database, New England Research Application Center (US), and Japan Light Metal Association.

WACK wait before transmitting positive acknowledgement.

WADEX word and author index.

WADS wide area data service. Data transmission service.

WAIT Western Australia Institute of Technology. Database originator and operator.

WAK 1. wait acknowledge. 2. write access key.

WAMI World Association of Medical Informatics.

WAN wide area network.

WAND Working Party on Access to the National Database. *CAG* study group on library computer communication (UK).

WAP work assignment procedure.

WARC World Administration Radio Conference.

WARC-ST *WARC* for Space Telecommunications.

WASAR wide application system adaptor.

WAT Wide Area Telecommunications Service, Bell System (US).

WATCON Waterloo concordance. Text handling package, University of Waterloo (Canada).

WATDOC *Water resources document reference system*. Database, Environment (Canada).

WATERLIT *Water literature*. Database, South African Water Information Centre.

WATFOR Waterloo FORTRAN. Version of *FORTRAN*, University of Waterloo (Canada).

WATS wide area telephone service.

WATSTOR *National water data storage and retrieval system*. Databank, *USGS*.

WB 1. Weber. Unit of magnetic flux. 2. wide band, telecommunications.

WBS wide band system, telecommunications.

WC word count.

WCB 1. way control block. 2. *Weekly criminal bulletin*. Database, Canada Law Book.

WCF workload control file.

WCGM writable character generation memory.

WCI waiting calls indicator, telecommunications.

WCL word control logic.

WCM writable control memory.

WCR word count register.

WCS 1. Wang Computer Systems. 2. writable control storage.

WCY World Communication Year. 1983 *UN* project to promote balanced development of information and telecommunication services.

WD 1. word. 2. write data. 3. write direct.

WDB *Werkstoffdatenbank*. Databank on properties of steel materials, *VDEh*.

WDB-1 *World data bank*. Databank of national borders expressed as coordinates, Harvard University (US).

WDC 1. Western Digital Corporation (US). 2. World Data Centre. Scientific data collection and exchange organization.

WDC-A *WDC* for US.

WDC-B *WDC* for USSR.

WDC-C *WDC* for Western Europe.

WDCS writable diagnostic control store.

WDPC Western Data Processing Centre, *UCLA* (US).

WE write enable.

WEP *World economic prospects*. Computer-based forecasting service, HM Treasury via Scicon Computer Services (UK).

WESLINK West Midlands Library and Information Service. Library cooperative (UK).

WESTAR west star. Internal communications satellite, Western Union (US).

WESTLAW West Publishing law system. Legal retrieval system (US).

WFL work-flow language.

WGN white Gaussian noise.

WHCLIS White House Conference on Library and Information Services 1979, Washington DC (US).

WHO World Health Organization. Databank originator.

WHP *World hydrocarbon program*. Databank, *SRI*.

WI The Welding Institute. Database originator and operator (UK).

WICAT World Institute for Computer-Assisted Teaching (US).

WICS Westinghouse integrated compiling system. Videodisc based system.

WILCO Wiltshire Libraries in Cooperation. Library cooperative (UK).

WISA Wormald International Sensory Aids. Manufacturer of aids for the handicapped.

WISE 1. Wang inter-system exchange. 2. World Information Systems Exchange. *IT* collaboration and exchange organization.

WISI world information system in informatics.

WKQDR work queue directory.

WL word line

WLC Wisconsin Library Consortium. Library cooperative (US).

WLN 1. Washington library network. Cooperative network (US). 2. Wiswesser line notation. Method for representation of chemical compounds.

WM word for meetings name. Searchable field, *OLS*.

WMI Wolfson Microelectronics Institute, Edinburgh University (UK).

WMO World Meteorological Office. Databank originator and operator (Switzerland).

WNI Windkracht Nederland Information Centre. Originator and its database and databank (Netherlands).

WO write only.

W/O without.

WOM 1. write-only memory. 2. write optional memory.

WP 1. word processing. 2. word processor. 3. write permit. 4. write protect.

WPDA writing push down acceptor.

WPI *World patents index*. Patent information database, Derwent Publications (UK).

WPM 1. words per minute. 2. write protect memory.

WPR write permit ring.

WPS word processing system.

WR working register.

WRAIS wide range analog input subsystem.

WRB Wissenschaftliches Rechenzentrum Berlin. Scientific computer centre (FRG).

WRC Water Research Centre. Database originator (UK).

WRENDA *World request for neutron data measurements*. Databank, *IAEA*.

WRIU write interface unit.

WRLIS Wessex Regional Library and Information Service. Medical library (UK).

WRSIC Water Resources Scientific Information Center. Database originator (US).

WRU who are you? Identify request, computing.

WS 1. working storage. 2. working space.

WSCA *World surface coatings abstracts.* Database, Azko Zout Chemie (Netherlands) and Paint Research Association (UK).

WSF work station facility.

WST word synchronizing track.

WT 1. witness terms. Searchable field, *SDC*. 2. word terminal. 3. work type. Searchable field, Dialog *IRS*.

WTA *World textile abstracts.* Database, Shirley Institute (UK).

WTI *World translations index.* Database compiled by the International Translation Centre, *CEC* and *CNRS*.

WTIC World Trade Information Center. Database originator (US).

WTS word terminal synchronous.

WU Western Union. Telecommunications company (US).

WUC Western Union Corporation (US).

WUIS *Research and technology work unit information system.* Databank, *DDC*.

WUTC Western Union Telegraph Company (US).

WW wire wrap.

WWD Weltwirtschaftsdatenbank. Originator and databank on business information (FRG).

WW MCCS Worldwide Military Command and Control System (US).

WWW *World weather watch.* Databank, *WMO*.

X

X 1. (as prefix) set of *CCITT* recommendations. 2. multiplication symbol.

XB crossbar switch.

XBC external block controller.

XBM extended *BASIC* mode, *ICL*.

XBT crossbar tandem. Electromechanical telephone exchange.

XCS Xerox Computer Services.

XCU crosspoint control unit, in telecommunications system.

XD ex-directory. Unlisted subscriber, telecommunications.

XDS Xerox Data Systems.

XDUP extended disc utilities program.

XFC 1. extended function code. 2. transferred charge call, telecommunications.

Xfer transfer.

XGP Xerox graphic printer.

XICS Xerox integrated composition system. Computer typesetting system.

XIO executive input/output, computing.

XIOP block-multiplexer input/output processor.

XIT extra input terminal.

XL 1. execution language. 2. (as prefix) distributed processing system, Pertec (US).

XL/OS *XL* operating system, Pertec (US).

XM expanded memory.

XMOS cross metal oxide semiconductor.

XMS Xerox memory system.

XMTR transmitter, not an acronym.

XNOS experimental network operating system.

XOP extended operation.

XOR exclusive *OR*. Logic operation or electronic gate.

XOS Xerox operating system.

XPD cross polarization discrimination. Antenna characteristic.

XPI cross-polarization interference, in radio transmission.

XPSW external processor status word.

XPT crosspoint. Switching element.

XR index register.

XRM external relational memory.

XS3 excess 3 code. Binary arithmetic transformation code.

XTAL crystal.

XTC external transmit clock.

XTEN Xerox Telecommunications Network. Terrestrial microwave user to user network (US).

Y

YA year authorized. Searchable field, Dialog *IRS*.

YAG yttrium aluminium garnet. Laser material.

YEC youngest empty cell.

YKB *Yukon bibliography*. Geographical database, University of Alberta (Canada).

Y/N yes/no. Response prompt.

YP year of publication. Searchable field, *NLM*.

YR year. Searchable field, *ESA-IRS* and Dialog.

YTD year to date.

Z

Z 1. impedance. 2. (as prefix) Zilog microprocessor.

ZA Zentralarchiv für Empirische Sozialforschung. Databank originator (FRG).

ZADI Zentralstelle für Agrardokumentation und Information. Databank originator (FRG).

ZAED Zentralstelle für Atomikenergie Dokumentation. Database originator (FRG).

ZAFO Zimmermanns Allgemeine Formenordnung. Zimmermann general form classification for engineering components.

ZAPP zero assignment parallel processor. For distributed computer processing.

ZBID zero bit insertion/deletion.

ZCR zero crossing rate.

ZDB *Zeitschriftendatenbank*. Machine readable database (FRG).

ZDE Zentralstelle Dokumentation Elektrotechnik. Originator and database on electrical engineering (FRG).

ZDS Zilog development system.

ZFM Zentralblatt für Mathematik. Database originator (FRG).

ZI zoom in, photography and video.

ZIF zero insertion force. Socket for *DIL IC* devices.

ZLC *Zinc, lead and cadmium abstracts*. Database, Zinc Development Association (UK).

ZOH zero order hold.

ZPB Z80A processor board, North Star Computers (US).

ZPID Zentralstelle für Psychologische Information und Dokumentation. Database operator (FRG).

ZRM zone reserved for memory. Olivetti (Italy).

217